# BIOLOGY Write NOW!

# BIOLOGY

# NOW!

T. L. TAIGEN
University of Connecticut, Storrs

J. MONNINGER
Plymouth State College

McGraw-Hill, Inc.
New York St. Louis San Francisco Auckland Bogotá Caracas Lisbon London Madrid Mexico Milan Montreal New Delhi Paris San Juan Singapore Sydney Tokyo Toronto

Biology Write Now!



1 2 3 4 5 6 7 8 9 0 MAL MAL 9 0 9 8 7 6 5 4 3 2

ISBN 0-07-003143-6

This book was set in Helvetica by Ted Taigen.
The editors were Mary Eshelman and Alice Mace Nakanishi.
The production supervisor was Pattie Myers.
Production assistance was provided by Jane Moorman.
Chapter illustrations were by Lad Hanka, Jane O'Donnell, Lisa Prince, Mary Jane Spring, and Libby Walker Davidson.
Malloy Lithographing, Inc., was the printer and binder.

Grateful acknowledgment is made for use of the following illustrations: *Pages 1, 21, 51, and 81* reprinted from *Animals: 1,419 Copyright-Free Illustrations of Mammals, Birds, Fish, Insects, etc.: A Pictorial Archive from Nineteenth-Century Sources,* selected by Jim Harter (New York: Dover Publications, Inc., 1979); *page 7* reprinted from *Life in the Cold: An Introduction to Winter Ecology* by Peter J. Marchand, illustration by Libby Walker Davidson, adapted with permission (Hanover, New Hampshire: University Press of New England, 1987).

# *Contents*

## About the Authors

**Theodore L. Taigen** is an associate professor of biology at the University of Connecticut. He holds B.S. and M.S. degrees from Colorado State University, and a Ph.D. from Cornell University. He has taught biology to science and nonscience majors, at the introductory and graduate levels. He has published numerous research papers in the field of animal physiological ecology and coauthored *Hands-On Biology*, a laboratory manual for nonscience majors. He grew up with Monninger, over whom he has enjoyed a lifetime of domination in basketball and pool. He would prefer not to discuss whiffle ball, darts, and Ping-Pong.

**Joseph Monninger** teaches English and writing at Plymouth State College. He holds a B.A. degree from Temple University and an M.A. from the University of New Hampshire. He has published six novels, including *Summer Hunt, The Family Man, New Jersey, Second Season, The Viper Tree,* and *Incident at Potter's Bridge*. He has also written numerous nonfiction articles for a wide variety of national magazines, including *Sports Illustrated, American Heritage,* and *Poets & Writers*. He has served as an editor for *Scientific American*. He grew up with Taigen and enjoys discussing whiffle ball, darts, and Ping-Pong.

# Preface

The good news is that colleges and universities across the country are reaffirming the importance of writing skills by requiring that students write "across the curriculum," as the latest wave of pedagogical newspeak grandly describes it. What could be better? Students will develop their communication skills in the courses closest to their hearts and minds—those that are central to their intellectual interests. No longer will writing instruction rest entirely on the shoulders of the English Department. Everybody will share in the joys and tribulations of helping students learn to write more effectively.

The bad news for a lot of biologists is that they now have to teach something with which they may not feel completely comfortable, using exercises that don't seem quite appropriate, to students who are somewhat skeptical about the whole process. The problem is that there are two basic plans for teaching writing in a science class, and both of these have flaws. Plan A is to have the students write up the results of their laboratory exercises, either as a detailed report, or more formally, as a manuscript. Plan B is to have them prepare a term paper. The frustrations with both approaches become apparent to anyone who has tried them. "Manuscripts" often become exercises in inflation, both verbal and intellectual, as students struggle to discuss results that are either awkward to explain if the experiment didn't work, or simple reaffirmations of textbook material if it did. Term papers are no better. Plagiarism becomes an enormous temptation for some students, fueled no doubt by the availability of mail order "Term Paper Assistance" companies which, for a fee (MasterCard and Visa accepted, of course), are willing to provide access to thousands of term papers. As the ads in the back pages of *National Lampoon* proclaim: If you've got the "term paper blues," help is only a phone call away!

This book is about alternatives, about Plans C and D, about ways to help students learn to write and at the same time teach them about biology. The conceptual organization of the book is apparent enough from the table of contents and the introductory comments that precede each unit. The first three units ask students to write personal essays that convey their views and opinions on topics ranging from the nature of scientific thought to the performance of steroid-dependent athletes. Our hope is that student effort will be inspired by focusing on subjects that are both topical and controversial. Ultimately, the success or failure of these essays depend on the degree of personal involvement by the student. The more we engage them in these topics, the more we ask for their personal interpretation of complex issues, then the greater will be their growth as writers and as independent intellects, capable of critically analyzing demanding problems for which there are no easy answers.

The ultimate writing challenge to students in a biology curriculum is the preparation of a scientific manuscript. The final unit instructs students in the entire process of preparing a manuscript, from formulating the initial hypothesis to dotting the final "i." The skills required to prepare such a document are the same as those developed in the essays that emerge from the preceding three units: critical thinking coupled with careful, precise use of language. In fact, we regard the scientific manuscript as the natural culmination of the intellectual process set in motion in the first chapter of this book.

We thank many people for their help in writing this book. Jan Taigen has contributed significantly to every aspect of development. Lisa Prince, Lad Hanka, Jane O'Donnell, Mary Jane Spring, and Libby Walker Davidson skillfully transformed our ideas into the illustrations and figures that grace each chapter. Robert Vinopal, R. Jack Schultz, and Tyler Taigen critically evaluated portions of the manuscript. Dotty Smarsik, Andy Moiseff, Tom Terry, Janine Caira, and Alice Mace Nakanishi helped with various aspects of formatting the manuscript. We have benefited from the inspiration, insights, and experience of our friends, graduate students, and undergraduates in the development of these ideas, especially Tom Polman, Steve Ressel, Judy Kucharski, Bill Mautz, Simone Nadeau, and Al Bennett. Finally, we express our deep appreciation for the many ways that our editor, Mary Eshelman, has contributed to the successful completion of this project.

# *To the Student*

So, you're in a biology class. Chances are you've already purchased a large textbook, maybe a very large textbook. And though it may be a bit daunting (in both length and weight!) it at least appears familiar. You've seen those kinds of textbooks before. You have also learned how to go through such a book, highlighting here, underlining there, taking detailed notes on items that will some day appear on an exam. To some degree, at any rate, you are comfortable with the way the material is presented and the demands imposed on you. It's what you anticipated.

*Biology Write Now!* is not exactly what you anticipated, we hope. For one thing, it looks different. It's only about one hundred pages long and it doesn't have any illustrations of the human reproductive system. Neither does it have a bunch of diagrams that break things into bite-size morsels. Instead, it was designed to be read in an easy chair. The chief difference between this book and a typical text is that it invites you not to memorize facts in preparation for an exam, but instead to participate in a noisy, dynamic dialogue about controversial issues and ideas in biology.

The way we want you to participate in this dialogue is through writing. Some of you may have come into your biology class believing that biology has little room for debate. It's a widespread misconception that science is black and white, and that scientific explanations are correct or incorrect with no use for a well-qualified "maybe." After all, isn't a scientist in a lab coat often used to symbolize rigid analysis, cold reasoning, absolute certainty? The popular myth insists that if all scientists are not like Mr. Spock from the planet Vulcan, motivated entirely by pure logic, then they should be.

In truth, the sciences are far more akin to a group of noisy relatives arguing around a kitchen table. Passionate, sometimes thick-headed, occasionally spiced with flashes of brilliant insight and blinding stupidity, science is an engaging and emotionally charged discussion that challenges, inspires, and frustrates, all at the same time. We hope this book captures that spirit of science. The topics we have selected for your appraisal are ones that have fascinated and puzzled us, ones that, in an intellectual sense, bring us to the "kitchen table," anxious to learn more, anxious to participate in the dialogue.

We have enjoyed writing the background material for each chapter and hope that it conveys to you our feelings of enthusiasm for these subjects.

Each unit will introduce you to a new theme, and within these units are chapters that will ask you to evaluate some background material, sort out your own analysis and opinion, and then write an essay that conveys your point of view. In a sense, your essays will not be "right" or "wrong" as, for example, your answers to a multiple-choice exam might be, but that is not to say that essays are indistinguishable beyond elements of style and prose. The chapters will invite you to think critically, and the best essays will be those that reflect careful analysis, include appropriate references to facts and reason, and demonstrate well-organized, efficient communication. For example, you may wish to argue that the use of data collected by the Nazis during World War II is appropriate (Chapter 2), or that the application of pesticides in ornamental plant cultivation is excessive and unnecessary (Chapter 13). Great, but you must be prepared to support your opinion. In some cases this will require you to dig deeper into the published literature to develop the evidence necessary for defending your view. Chapter 16, "Using and Abusing the Scientific Literature," will help you do that. In all cases, you will want to convey your ideas as efficiently and clearly as possible. Chapter 18, "Effective Scientific Writing: Style and Format," will give you some ideas on how to accomplish that. Both of these chapters should serve as resources for improving your essays.

Now on to a short talk about writing.

## How Do I Know What the Instructor Wants?

Your instructor, any instructor, wants good writing. Here are some tips to help you improve your essays.

1. Read the introduction for each unit. These comments will give you an overview of the material, and help you to appreciate the broader philosophical perspective of the chapters that follow. They will also prepare you for the ensuing writing exercises.

2. Start early. Get your essay under way as soon as possible. Brainstorm ideas, jot down tangents of thought, but get going. Put it on a word processor so that monkeying around with it will be easier. The sooner

you achieve a first draft, the more time you can devote to proofreading and revising.

3. Write to illuminate, not to confuse. Try to imagine how your prose appears to someone reading your work. When you have finished a draft, read the essay aloud. By doing so, you'll hear if it makes sense, reads smoothly, and so on. Then ask a friend to read it. Make sure the friend understands that you want to be challenged on the content and form, not merely slapped on the back and told it's great. Are the claims and opinions you present in your essay fully supported?

4. Be concise and stick to the point of the essay. Many essays are undone by a rambling, unfocused string of paragraphs, the point of which is never very clear. In the margin of your paper, write down the central thought of each paragraph. Do the paragraphs lead gracefully from one to the next? Is the flow of ideas logical? Is your conclusion clearly developed from the preceding paragraphs?

5. Avoid rhetorical questions in your essay. These may work for Johnny Carson during his monologue, but they are rarely effective in scientific expository writing.

6. Proofread your work before turning it in. Perhaps regrettably, spelling and grammar always "count," whether you realize it or not. Make sure you are writing in complete sentences. How do you know? Check for three things: subject, verb, and complete thought. You might also want to pull your essay apart and examine each sentence as though it were an example in a grammar book. Write it down line by line and look at it. It's remarkable how clear a sentence becomes (or doesn't) when pulled away from the surrounding material.

## May I Use the First Person?

The first person "I" is extremely effective at times. Be aware, however, of the tendency to become more subjective and less objective when using the first person. Subjectivity in itself is not a fault; unreasoned subjectivity is. Claims and interpretations must be supported by fact, not impressions.

The most effective means of using the "I" is in anecdote. The following introductory paragraph was written by a student responding to Chapter 13, "Dandelions Versus the Suburban Turf Warrior." Notice how she pulls the reader into the situation and heightens the sense anticipation for the material to follow through her use of the personal pronoun.

> "The foul smell of chlordane is fixed in my nose. As a child, I learned quickly to squash the parade of black carpenter ants that invaded our kitchen every summer. It was never long, though, before my father discovered the little creatures marching authoritatively across the Formica counter and, before the explicative had left his lips, coupled the jar of chlordane to the garden hose and sprayed the foundation with a vengence. The stench was overpowering and lingered for days. My father now accepts the possibility that chlordane is hazardous and has, perhaps, been banned with good reason. But he still hates ants in the kitchen, though, and this year came home with a very large bag of diazinon and a new smell permeated the house—the battle continues."

By using the first person she has given us someone with whom we can identify. The argument is no longer abstract. One of the most respected contemporary science writers, Stephen Jay Gould, almost always uses his personal reflections before digging into the meat of the issue. If you'd like to sample his work, pick up a few copies of *Natural History* magazine in your library and read his column.

## How Extensively Should I Use Outside Sources?

Each chapter lists additional references that will help you to learn more about the topic. In addition to these, try to discover other sources, especially those outside the published scientific literature. *Nova* programs produced by the Public Broadcasting Service (PBS) are usually beautifully done and may have information directly relevant to your essay. The science section of your local newspaper is another good place to look for current information. Use your imagination, or talk to your instructor, to come up with alternative references—the B'nai B'rith for the Holocaust essay, the National Institutes of Health (NIH) Office for Scientific Integrity for Chapter 3, or maybe People for the Ethical Treatment of Animals (PETA) for Chapter 9.

Outside references will help you develop your ideas, but they will not replace your judgment. Use the background material to support the claims and opinions you develop in your essay. Without supporting evidence, from either external references or the persuasion of your well-informed critical analysis, your opinion, as they say, is hardly worth the paper it's written on.

## How Long Should the Essay Be?

It's difficult to tell another person precisely how long an essay should be, because that infers there is a correct or necessary length required to cover an issue. From our experience with these topics, though, we've found that most of them can be adequately addressed in four to six pages of double-spaced text. Longer essays may result from a deeper analysis, but shorter pieces are almost invariably deficient in content or clarity. Your instructor may provide additional guidelines in this regard.

## Closing Remarks

Read. Think. Write. We have deliberately constructed these essays so that they are incomplete. We did not want to write the essay for you, or trample on whatever ideas you may hatch. Our chapters are intended as starting points, nothing more. You will soon realize that there is a substantial body of knowledge and information for you to explore. In rummaging through the material, be skeptical, but at the same time keep an open mind. Don't accept easy answers or jump too quickly to a conclusion. Don't defend a position blindly. Discuss, debate, read, research, then discuss again—with your friends, your instructor, your family. Try your intellectual wings, and don't be afraid to fall. The only way you can truly fall is by failing to try.

# *Unit I The Nature of Science*

The first chapter of this unit, "Science As a Way of Knowing," actually took its first incarnation as a question on a multiple-choice exam in a class one of the authors taught a couple of years ago. The question required students to recognize that science is but one of many "ways of knowing" that people use to understand themselves and their world. One student in the class, annoyed at having missed the question (or perhaps more accurately, at having to deal with an incalcitrant, pig-headed instructor who seemed intent on ignoring even the most coherent argument) asked for the chance to write an essay in which her view could be fully (and passionately!) articulated. The essay was a good one, clear, concise, compelling, and, claims of pig-headedness notwithstanding, sufficiently convincing to cause a re-evaluation of her exam.

From this lively (and somewhat confrontational) beginning, we have struggled to find ways to bring writing exercises into a biology curriculum. Our collective experiences, based on a multitude of topics and an even greater assembly of students, inevitably bring us back to the lesson of that very first essay. Students must truly care about an essay topic before they will tackle it with energy and enthusiasm. And the more deeply they care, the more powerful is the learning experience, the more effective is the written communication.

Above all, we want you to care about the issues raised in the chapters of this unit because in many ways they are the most important lessons you will learn in this course. This unit presents four essays dealing with different aspects of the process of science. We hope these essays will cause you to think about science in ways you never have before. They concern the human elements of science and, as such, involve questions of morality, integrity, motivation, and achievement. As you read these essays, you will probably note the repeated use of the expression, "There are no easy answers to these questions," and that admonition bears repeating here. But it is because there are no easy answers that these issues are such fertile ground for your intellectual efforts.

Each of the chapters presents some background information on the topic, followed by a series of questions that are intended to get you thinking. It isn't necessary for you to answer all the questions. In fact, in some cases, it isn't possible for you to answer all the questions (assuming that each essay will consist of four to six pages of typed, double-spaced text). For many of the questions, there are no "right" or "wrong" answers, there are only responses that do or do not reflect careful analysis, include appropriate references, and demonstrate well-organized, efficient communication. The best essays are those that go beyond the dimensions of the material presented in the chapters, either by the force of your own critical thinking about the issues, or by reference to literature and resources other than those cited at the end of the chapters.

Finally, we hope that you will read each of the chapters, even if only one or two of them are actually assigned as a writing exercise. If nothing else, we hope that the ideas presented in this unit will serve to reaffirm science not as an assemblage of facts and information, but rather as a dynamic, powerful, exciting, and, perhaps most significantly, profoundly human endeavor.

# *Science As a Way of Knowing* 1

*"You know better than that!"* When your parents told you that as a small child (and perhaps even still!), you may not have fully appreciated exactly how you were supposed to "know better," but there was no ambiguity whatsoever regarding their unfulfilled expectations or their evaluation of your behavior. Depending upon the seriousness of the transgression, that childhood admonition may have been followed by a parental expression of the causes (usually personal, rarely flattering) and consequences (inevitably dire) of the misjudgment in question. Underlying this entire exchange is an assumption

that the collective effect of years of upbringing in a particular cultural environment has instilled a "way of knowing" that will help you understand and deal constructively with problems you encounter in your life.

In some respects, a scientist struggling to understand the complex laws of nature is not really so different from a child trying to comprehend a bewildering and complicated world. The only real distinction, but one that makes all the difference, is that problem-solving for scientists involves reliance on a process, a "way of knowing," that is without equal for its usefulness in understanding the natural world. The scientific method, despite the uninspired (if not downright boring) cookbook descriptions applied to it in many introductory biology textbooks, is a powerful, sometimes frustrating, often tentative, always dynamic tool not only for learning about the world but also for solving many of the problems confronting it.

Although science can be viewed as a "way of knowing," the example given in the opening paragraph of this chapter suggests that it's not the only way of knowing and in some circumstances it may not even be the best. Politics, religion, art, mysticism, humanitarianism, even astrology and intuition offer approaches and methods that advocates use to understand themselves and their world. For some of these approaches, the underlying methods have a structure that is both clear and uncompromising. For others, the methods may be diffuse and vague, though no less effective in helping make decisions or deal with problems.

In a world that seems to spin almost instantaneously from one dramatic new development in human affairs to the next, sometimes it's hard to tell the difference between science and other ways of knowing. Sometimes even scientists can't tell. The complexity and pace of our world can blur the distinction not only between science and pseudoscience, but also between religion and mysticism, between politics and humanitarianism, between astronomy and astrology. Your assignment is to write an essay comparing science as a way of knowing with other ways of knowing. What are the advantages and disadvantages of the various methods? How do the principles upon which they are based differ? Do you agree that they can all be described as "ways of knowing," or are there differences among them so fundamental as to render superficial any intellectual scheme that categorizes them as simply variations on a theme? What types of questions, or subject matters, does each way of knowing address? In what ways are they limited? Finally, if these ways of knowing differ in their strengths and weaknesses, what are some of the dangers (or cautions) inherent in the strengths and what are the consequences of the weaknesses?

## Getting Started

The best way to write this essay is by starting with a precise description of the scientific method. Clearly and succinctly state how science proceeds and how it is limited. Then discuss the underlying method of other ways of knowing. When you are writing this part of your essay, be sure to develop clear distinctions between unquestioned truths and falsifiable hypotheses, between theory and dogma, between faith and critical thinking. Use examples where appropriate to make your point. It isn't necessary to include all the ways of knowing listed above to write an effective essay. In fact, you may want to limit your comparison to include only religion, or perhaps art, or politics. The important point is that you develop a clear and unambiguous statement of exactly how science differs from these other ways of knowing. After making that distinction, you should address the question of the where science overlaps with other areas, and where it does not. For example, is there a role for religion in science, and for science in religion? You should also deal with the idea that perceived limitations on the realm of science may very well reflect cultural or philosophical biases. Some theologians may appreciate fully scientific achievements in physics and chemistry, but deny entirely a role for scientific methodology in questions of human ancestry or the origin of life. Is this any different than disallowing a role for science in the fields of art or history?

You may wish to develop your essay by creating an example of an event that can be viewed from the vantage of several ways of knowing—a supernova perhaps, or maybe the extinction of the dinosaurs. How do the various ways of knowing deal with this event?

## References

In developing this essay, it may occur to you that some familiar words suddenly seem a bit slippery and uncertain. Precise use of words such as "knowing," "understanding," "truth," "wisdom," "theory," and "dogma" is a critical component of your essay. Use a dictionary to provide definitions of these words, and then apply the definitions in your discussion of the scientific method. These concepts are also developed in a good introductory philosophy textbook. They are usually organized under the topic of "Epistemology." (Use your dictionary first to look up "epistemology"!) Finally, the success of your essay turns on your ability to communicate the details of the scientific method. Use a good introductory biology textbook to clarify your understanding of the process of science.

# *Searching for Scientific Morality in the Ashes of the Holocaust* 2

Among the hideous atrocities committed by the Nazis during the Holocaust, perhaps none were more horrifying than the unspeakable acts performed in the name of scientific experimentation. While Nazi scientists observed and recorded data, human beings were subjected to brutal and degrading treatments, including fatal exposure to extreme temperatures and enforced sleeplessness to the point of exhaustion and death. Fifty years later, the criminals who perpetrated these acts are dead, but the data that they collected remain as part of the Nazi legacy. In 1988 scientists working for the

Environmental Protection Agency (EPA) suggested that some of these data had the potential to contribute important information regarding the effect of sublethal doses of toxic gases on workers involved with the manufacture of these materials. Obviously, the kinds of experiments necessary to provide that type of information can never be repeated, and some scientists feel that using the data to improve the human condition in effect salvages something good from the blackest chapter in the history of human activity. For many scientists, this situation imposes a difficult and delicate moral dilemma for which there are no easy answers.

It can be argued, for example, that to allow anything "good" to emerge from the Holocaust is a profanity, a desecration of the lives of those who perished and an insult to those who survived. The final and most demanding lesson of that period of human history is that it must never be repeated. To allow that lesson to become diffused by "salvaging something good" is tantamount to softening the face of an evil so terrifying and so destructive that it can never be allowed to be anything other than the final expression of human hatred and savagery. Even if a cure for the most deadly form of cancer lies among the ruins of Nazi science, it should remain buried there forever because the integrity of that lesson must never be compromised.

These arguments make a powerful appeal to a sense of history and the sanctity of human suffering, but one of the inescapable consequences of this rigid noncompromising position is that people may suffer or even die as a result. Even though the circumstances under which the data were collected are reprehensible, the simple fact remains that this information could save lives and help people *now*. Does it diminish the lessons of the past or the lives of those who suffered to use the data in a constructive way? How do you explain to a worker who has suffered a lifelong, debilitating injury as a result of prolonged exposure to sublethal concentrations of a toxic gas that his condition could have been avoided if the EPA had adopted a less strident approach to evaluating Nazi science? And might your reaction to this question be different if you or someone you love were afflicted by a disease that could be cured using information that is available, but deemed "untouchable"? As in every other human endeavor, personal circumstances have an uncomfortable and disconcerting way of restructuring moral interpretations of the scientific process.

At the core of this moral conundrum is the fundamental question of what is morality and how does it apply to science. This dilemma is hardly limited to questions about the legitimacy of Nazi science; in fact, it occurs constantly in the scientific community. For example, a long-running debate in the field of neurobiology centers on the phenomenon of pain and whether or not the mechanics of this phenomenon should be studied. Obviously, the potential payoffs for that type of research are enormous. An understanding of how pain is perceived and processed may lead to wholesale alleviation of human suffering. And yet, how can researchers study pain without evoking that sensation in their experimental subjects, either human or animal? Is it morally legitimate to cause pain and suffering in animals in the hope of learning how to eliminate pain and suffering from the lives of billions of human beings?

These questions have no easy answers. The essay you write will explore the boundaries of morality in science. Are there questions in science that should never be asked because of moral constraints? If so, who should decide on such constraints—a larger society? Individual scientists? Or perhaps there is a morality that is intrinsic and unique to science which should be the guide to questions such as these. Finally, what should be the dimensions of moral and ethical considerations in science? To what extent should scientists be told what they can and cannot do? And what should we do with the ashes of the Holocaust?

## GETTING STARTED

This is not an assignment that can be met merely by organizing and paraphrasing what has been written by others. The topic calls for your critical evaluation and thoughtful consideration, and it is *your* opinion that is important here, not that of some other scientist (or more likely professional science writer) who has written a clever, easily paraphrased (or worse, plagiarized) essay on a narrowly defined topic. As you write your essay, be aware of the questions that have been posed above. These are intended to get you thinking, but you are not required to provide an answer for each of them. As suggested in the unit introduction, there are no "right" and "wrong" answers to these questions. The strength of your essay will depend on the depth of your critical analysis, the breadth of your references, and the effectiveness of your writing style.

At the minimum, your essay must do the following: It must unambiguously define your view on morality in science, and then apply that definition to the specific question of the Nazi data. In order to make a strong case for your judgment, you should be prepared to support your position by using specific examples. What types of research should never be conducted? Give examples and explain why. Certain animal rights groups, for example, might argue that any scientific research involving the destructive use of animals is immoral, and some theologians contend that experiments conducted on fertilized human embryos in a petri dish are a violation of the sanctity of human life. Once you have identified research that falls outside the moral boundaries of science, you must clearly communicate why this is so.

Finally, be aware of the distinction between sanctioning a particular type of scientific research and simply making use of the resulting data. The real question is, can the results produced by immoral research ever be extracted from the dark context in which they were produced and allowed to become "useful"?

## REFERENCES

The references given below are intended to help you learn more about the background material presented at the beginning of the chapter. In addition to these, your instructor can help you to find material relevant to the essay you write.

Astor, Gerald. 1985. *The 'Last' Nazi: The Life and Times of Joseph Mengele.* New York: Donald I. Fine.

Feig, Konnlyn G. 1979. *Hitler's Death Camps: The Sanity of Madness.* New York: Macmillan Publishing Co., Inc.

Gray, Bradford H. 1975. *Human Subjects in Medical Experimentation.* New York: John Wiley and Sons.

Lifton, Robert J. 1926. *The Nazi Doctors—Medical Killing and the Psychology of Genocide.* New York: Basic Books, Inc.

Shabecoff, Phillip. 1988. Head of EPA bans Nazi data in study on gas. *The New York Times.* March 23, pp. A1 and A17.

Sun, Majorie. 1988. EPA bans use of Nazi data. *Science* 240:21.

# *The Black Sox of Biology: Trofim Lysenko and the Shattered Integrity of Soviet Genetics* 3

He cheated, he lied, he violated every ethical standard in science, and he set in motion a scientific and economic catastrophe that still reverberates through the chaotic world of the Soviet Union today. For nearly 30 years, Trofim Lysenko and his bizarre genetic theories dominated Russian science, isolating it from the rest of the scientific world and destroying a distinguished tradition of achievement established by a generation of Russian geneticists. It is difficult to imagine, in today's world and with the crystal clarity of hindsight, how such events could have occurred. In fact, Lysenko's climb to power, literally over the bones of courageous Russian scientists who were executed for their commitment to scientific integrity, began and ended with his perception that scientific hypotheses must foremost reaffirm the Marxist-Leninist philosophy espoused by the totalitarian ruler, Joseph Stalin. Over a period that began in the late 1930s and persisted until the bankruptcy of Lysenko's ideas were fully revealed in 1965, genetic theories in the Soviet Union were evaluated not on the basis of supporting evidence, but rather on the extent to which they were "politically correct."

The absurdity of Lysenko's theories are now fully appreciated. He vigorously denied the existence of genes, the importance of chromosomes, and the principles of Mendelian inheritance. He refused to accept the idea of natural selection, referring to it as "Darwin's mistake," and unconditionally embraced a neo-Lamarckian view of inheritance of acquired characteristics. Lysenko passionately believed, inspired by Marxist philosophy, that organisms adapt to their environment and subsequently pass on such "acquired characters" to their offspring. Perhaps most the most strikingly unconnected of Lysenko's assertions was his belief in spontaneous generation and the instantaneous transformation (called "transmutation") of species from one form to another. That such ideas were championed by the director of the Soviet Genetics Institute is frightening; that they served for decades as the basis for plant and animal breeding programs in Russian agriculture is both terrifying and, as history has shown, a sure recipe for disaster.

As far as can be determined, Lysenko never actually experimentally tested any of his ideas, although his supporters would fabricate clumsy falsifications of data to support his officially sanctioned dogma. For example, followers of Lysenko once published a photograph in a Soviet botanical journal that purported to demonstrate the transmutation of pine to birch in a forest near Leningrad. The photograph convincingly showed the lower half of the tree to be a pine, the upper half a birch. Years later, after Lysenko's fall from political grace, a new photograph, taken from a different angle, was published in the same journal, clearly showing two entwined, partially fused trees that were simply growing very close together. Lysenko's supporters, Stalin first among them, may have been many things—brutal, dogmatic, politically powerful—but sophisticated they were not.

The disastrous consequences of "Lysenkoism" extended from destruction of individual careers to undermining the national economy of the USSR. Those scientists who opposed Lysenko were labeled dangerous bourgeois reactionaries, inventing "myths" to advance their counterrevolutionary goals. Many geneticists, who's only "crime" was that they would not renounce the gene as the unit of heredity, were declared to be scientific enemies of the state and either executed directly or sentenced to years of degradation in Stalin's torture chambers. The victory of Lysenko's doctrine was so complete that modern genetics was banned from the curriculum of Soviet universities for decades. Without a legitimate scientific base, agricultural technology in Russia foundered. Only after years of complete failure (and, perhaps more significantly, the death of Stalin) did the scientific community haltingly begin to criticize the plant breeding program directed by Lysenko. Unfortunately, by that time, the national economy of the USSR had listed into a malaise of nonproductivity from which it has never fully recovered.

The tragic history of Trofim Lysenko serves as a compelling reminder of the vulnerability of the scientific process. Lysenko used the full persuasion of politics and philosophy to insert his dogma in place of legitimate science. It may be argued that what he did is simply an extreme variation on a recurring theme—science can be biased by the political, philosophical, and cultural perspectives of the scientist. Perhaps regrettably, scientists are no different from anybody else in society; they have and respond to biases and prejudices. Your assignment is to write an essay that explores the relationship between cultural bias and the scientific process. The role of race, gender, politics, and public policy in the scientific process should be addressed in your essay. The purpose of the essay is to express your understanding and appreciation of the idea of scientific integrity. To what extent is science "above" the cultural biases that infiltrate so deeply into human activities? Can immunity from those biases ever be achieved in the scientific process? And if it can, is this necessarily a positive development for scientific progress? In what kinds of circumstances might cultural biases actually help in learning more about the natural world?

## Getting Started

A good way to start this essay is to ask yourself if it is possible for a Lysenko to exist in today's scientific world. The keys to understanding Lysenko's rapid ascent and long reign are that his ideas were politically correct and that he used the popular press instead of the scientific literature to communicate his views. Can this sort of thing occur in this country? A couple of years ago, a news conference was called at the University of Utah to announce the discovery of "cold fusion." The news release to the popular press created a great furor in the scientific community because it preceded publication of the data in a peer-reviewed scientific journal. Does this situation bear analogy to that of Lysenko and Soviet genetics? What are the dangers of using the popular press to communicate scientific information? How important is the scientific literature for maintaining integrity in science?

In the end, Lysenko's doctrine was clearly seen as state-imposed dogma, not science at all. Is it possible for this type of event to occur now? Over the past several years, fundamental religious groups have attempted to insert "creation science" into the biology curricula of public schools, to be taught side by side with evolution. The claim is that the theory of evolution is no more or less "scientific" than the theory of creation by divine intervention. Do you agree? Again, does this situation bear analogy to that of Lysenko?

You may wish to focus your essay on more subtle ways in which science reflects cultural and political biases. Go to the science section of a newspaper in your area and read the articles from this perspective. Is there evidence of bias? How might the research have been different if the gender or race of the scientist was different? At what point in the scientific process do cultural biases come into play? The design of experiments? The analysis and interpretation of results? Or perhaps biases are only evident in the kinds of questions the researcher is capable of asking about a phenomenon under study. Some might argue the real question is not whether science is biased, but how. Do you agree? Use the information from the articles to support your position on the integrity of science.

## References

Berg, R. 1978. *Acquired Traits: Memoirs of a Geneticist from the Soviet Union.* New York: Viking-Penguin.

Gershenson, S. M. 1990. The grim heritage of Lysenkoism: Four personal accounts IV. Difficult years in Soviet genetics. *Quarterly Review of Biology* 65:447–456.

Joravsky, D. 1970. *The Lysenko Affair.* Chicago: University of Chicago Press. (Also the basis of a PBS *Nova* production [1974] with the same title)

# *Looking Up in Perfect Silence*

# 4

## WHEN I HEARD THE LEARN'D ASTRONOMER

by Walt Whitman

When I heard the learn'd astronomer,
When the proofs, the figures, were ranged in columns before me,
When I was shown the charts and diagrams, to add,
    divide, and measure them,
When I sitting heard the astronomer where he lectured with much
    applause in the lecture-room,
How soon unaccountable I became tired and sick,
Till rising and gliding out I wander'd off by myself,
In the mystical moist night-air, and from time to time,
Look'd up in perfect silence at the stars.

"When I Heard the Learn'd Astronomer" is a famous poem by one of America's most popular poets. It was written in the middle of the nineteenth century, at about the time of the Civil War. At that point in our history, educational evening lectures were quite common. Well-known scientists and writers (as well as opera companies and symphony orchestras) visited even the smallest towns to speak on various topics. Mark Twain was extremely well known for this type of performance, and Charles Dickens, visiting this country from his native England, gave such brilliant performances that women often fainted in the audience. On the night described in Whitman's poem the lecturer is apparently articulate and well prepared, although no one appears on the verge of fainting. He provides diagrams and measurements, instructs the audience on the available facts and figures, and is rewarded for his efforts by rousing applause. In short, it's a good lecture. Despite the abilities of the orator, Whitman goes on to tell us, "How soon unaccountable I became tired and sick/ Till rising and gliding out I wander'd off by myself/ In the mystical moist night-air, and from time to time/ Look'd up in perfect silence at the stars."

The poem is a favorite among English teachers. Not only is it representative of Whitman's work, it also makes a compelling point about beauty and mystery in the perception of nature. Whitman reminds us that the wonder of the night sky cannot be relegated to triangulation and celestial cartography, that a detailed structural analysis of the stars and the moon and the clouds is not necessary to appreciate the beauty of the heavens. In fact, there is even a suggestion that scientists, armed with their precise methodology and quantitative rigor, are the ones left behind in the lecture hall. They sit happy and pleased with their narrow contemplation of numbers and figures while the true wonder of nature quietly slips by outside the door, noticed only by those whose imaginations are unencumbered by the straitjacket of "scientific thinking." Unfortunately, that view of science clashes with the way most scientists perceive themselves. Beauty and elegance and mystery are regarded by scientists as catalysts of the scientific process, not victims. Although their methods are different, scientists like to think of themselves as standing side by side with poets and painters, fascinated by the world around them, searching it for greater mysteries and deeper beauty.

Perhaps more than anything, these reflections lead to an appreciation of science as a profoundly human enterprise, invested with all the advantages and drawbacks, the strengths and weaknesses, the full range of success and failure, implied by that characterization. Your assignment is to write an essay that addresses the question of motivation and achievement in science. The goal of the exercise is to explore the personal side of the scientific process, in contrast to the dry, cookbook, plug-in-the-question and crank-out-the-answer impression of science presented in introductory biology books. What drives scientists to do what they do? Material rewards? (This will be perceived as a joke by most scientists!) A quest for beauty? Intellectual curiosity? A search for order or symmetry? A love for a specific animal (e.g., whales) or plant or natural phenomenon? Or perhaps negative motivations are at work—scientists are not so much drawn to science as they are driven from more socially interactive pursuits. Perhaps scientists don't enjoy interacting with other people, preferring instead to interact (or maybe even compete) with nature. There is, of course, no single answer to this question, just as there is no one answer to what motivates other human endeavors. Each scientist represents a unique mosaic of motivations and personality. Your essay should deal with the ways in which these motivations are unique to scientists, how they may or may not differ from those of a sculptor, or an athlete, or a physician, or an engineer.

Your essay should also address the issue of the extent to which scientific orientations and perspectives spill into aspects of personal life. Do scientists evaluate their emotional, philosophical, and spiritual experiences in the same way as they do their experimental results? The issues here are deep and highly personal. Is it possible, for example, to reconcile the theory of evolution or the geological study of the age of the earth with the book of Genesis? How is it possible that Einstein, among the greatest scientific minds of this century, was a devoutly religious man? Is there room for the face of God in a mind that reiterates constantly the words of Sergeant Joe Friday, "Just the facts, ma'am." In a broader sense, at what point does scientific inquiry limit or intrude upon human imagination and spirituality? Does it deaden our appreciation of the world around us? Or, on the contrary, does it stimulate us toward greater comprehension and greater leaps of imagination? Do you find religion and science irreconcilable opponents? Has your increased understanding of how organisms work enhanced your appreciation for life? Do you find too much science a hindrance in trying to understand the universe, as Whitman apparently did? Or do you feel that just as poetry is a way of speculating about the nature of the world, science in its way can be every bit as poetic?

Finally, you may wish to deal with these questions in the context of an understanding of how progress is achieved in science. What is the source of great scientific insight? What circumstances or common experiences are associated with the moment of discovery? What sets the stage for success in science, and is that really different from successful endeavors in other human activities?

## Getting Started

The most direct way to deal with these questions and write this essay is to approach scientists and talk to them about their work. Why do they do what they do? Are there "personality types" predisposed to careers in science? How do they make progress in their research? Do they always adhere rigorously to the scientific method as it is described in your textbook? How or when do they deviate? What role does luck, both good and bad, play in scientific achievement? Ask about the importance of personality in scientific success. It has been argued that "achievement" is sometimes a complex result of good science coupled with good marketing skills, and in some cases it is difficult to tell which of these is more important. Wherever possible during your interview, try to elicit specific examples that you can draw upon in your essay. Some of these examples may require additional background research in the scientific literature.

After you have talked with one (or better, several) scientists, approach nonscientists and ask them the same questions. What do they think of scientists? How do they feel scientists differ from artists or athletes or engineers? How are they the same? Ask them about the reconciliation of spiritual experiences with scientific methodology. When you have finished your interviews, note the differences between the way scientists perceive themselves and how they are perceived by others. How can you account for the differences?

And finally, when you have put all this material together, when you have asked the questions and reflected on the answers, please decide exactly who it was that stood outside the lecture hall, looking up in perfect silence.

## References

Before you conduct the interviews, you should prepare yourself by reflecting carefully on the introductory essay in this chapter. In addition, you may wish to consult the references given below or other books on the philosophy of science. These will help you organize your interview and focus your essay.

Chandrasekhar, S. 1987. *Truth and Beauty: Aesthetics and Motivations in Science.* Chicago: University of Chicago Press.

Horgan, J. 1991. Profile: Reluctant revolutionary. *Scientific American* 264:40–49.

Judson, H. F. 1980. *The Search for Solutions.* New York: Holt, Rinehart, and Winston.

Kuhn, T. S. 1970. *The Structure of Scientific Revolutions,* 2d. ed. Chicago: University of Chicago Press.

Whitman, W. 1981. *Leaves of Grass.* New York: Random House, Inc.

# *Unit II New Perspectives on the World Around Us*

This unit is about new ideas. That in itself is hardly striking. After all, every issue of every scientific journal is filled with new ideas. What's different about the ideas in this unit is that they represent novel approaches to interpreting the world around us. When you finish reading the chapters in this unit, we hope you never again think about living organisms in quite the same way you did before. In separate chapters, organisms are described variously as energy transformers, gene machines, composites of functional and nonfunctional traits, pieces of a large and delicate ecological puzzle, and endowed with legitimate but conflicting rights. Some of these perspectives may strike a familiar chord with you (e.g., ecological puzzles and animal rights), but we hope you will find that the chapters deal with this material in a way different from any you have seen before.

As with many profoundly different ideas, some of these new perspectives have engendered a great deal of criticism and debate, both in the scientific literature and in the public media. We would like you to evaluate these perspectives, analyze the world from their vantage, then decide for yourself their strengths and weaknesses, their legitimacy or irrelevance, and finally, their usefulness for developing new insights about the natural world. You may be tempted to think that the authors or your instructor hope you will agree with one or all of these new views. If you think that, we hope you'll do exactly the opposite. Disagree. Trash it. Argue a dissenting position. The more intellectual energy and passion you bring to your essays, the more satisfaction we will enjoy at having done our job of helping you to care about what you write. We want you personally involved with the ideas presented here.

These are not term papers. Your instructor is not interested in reading a stack of loosely paraphrased restatements of material he or she has read, taught, and discussed for years. But he/she is very much interested in finding you right in the middle of it, up to your neck in the intellectual struggle of sorting out the ideas, the data, and most importantly, your own judgments. We like to think of every chapter in this unit as having two parts. One part is available to you in the literature and the background material; the other is available only in the depth of your personal reactions and the strength of your convictions. As the old recruitment slogan goes, "We want *you*!"

# *Piebald Foxes, Six-Toed Dogs, and the Evolution of Nonfunctional Traits* 5

For this essay, you're going to need a dog. It doesn't matter if it's yours, your friend's, your parents', whether it's a purebred or a mutt—just find a dog and take a good look at it. Take your time. Examine it. Then, with pen and paper in hand, make some notations about how the features of the animal you are looking at are different from those one might find in the wild ancestor of domestic dogs, the wolf. How do the ears compare? The tail? The length of the legs? The size and body conformation? The length and color of the fur?

When you have finished listing all the differences you can think of, consider the question of how these differences came to be. Specifically, what process or mechanisms led to the striking differences between the dog in front of you and its distant cousin roaming through the northern boreal forests? The purpose of this chapter is to explore the answers to that question and then to evaluate the extent to which they may occur in the evolution of natural populations.

One of the mechanisms, of course, you already know. For hundreds, in some cases thousands, of years, breeders have selected for specific traits considered desirable in a particular breed—strong jaws in pit bulls, for example, or diminutive size in a Chihuahua, or short legs in a dachshund. Through selective breeding, desirable traits were enhanced while those deemed undesirable were eliminated. Features are selected in these breeding programs either because of their functional significance (e.g., the aforementioned jaws of the pit bull) or on aesthetic grounds (the size of the Chihuahua). Thus, completely nonfunctional traits may appear in domestic breeds simply because they are perceived as aesthetically appealing.

In nature, aesthetics don't usually for count for much. Function, on the other hand, is a major, some would say exclusive, focus of natural selection. A trait is defined as "adaptive" in an evolutionary sense, when it fulfills a function that is critical to the animal's survival and reproductive effort. Thus, if the evolutionary process constantly selects for adaptive features that allow for a good fit between an organism and it environment, then that would seem to leave little opportunity for the appearance of nonfunctional traits in organisms. In fact such "opportunities" may exist, but perhaps not in ways that you would expect.

Consider the intriguing case of an experiment on domestication of wild foxes. For several decades, foxes were selectively bred on the basis of their reaction to humans. In the initial stages of the experiment, animals that displayed tame, nonaggressive behavior toward their handlers were chosen for breeding with individuals that exhibited similar dispositions. Later, as the animals became domesticated, only those individuals that were actively willing to contact their handlers physically were selected to continue in the breeding program. Over a period covering many generations, the collective changes in behavior and disposition were dramatic; foxes not only lost their fear of humans, but they had acquired many of the qualities typically found in the family dog, including whining for attention, tail wagging to communicate affection (even toward strangers), answering to nicknames, licking the hands and face of familiar handlers, and constantly searching for physical contact with humans. Anyone who has seen the shadow of a wild fox darting frantically across a road on a moonlit night or through a meadow just before dawn—every muscle fiber and every avenue of sensory perception fully engaged to detect and avoid human contact—realizes how remarkable are the results of this breeding experiment.

The tameness and emotional state of these foxes resulted not from training or repetitious exposure to humans, but rather from selection for specific genetically inherited traits. Along with the tame disposition, these animals also possessed a suite of unexpected and unique physiological and morphological traits. Reproductive cycles of tame females exhibited a number of abnormalities, the most unusual of which was the occurrence of "extra-seasonal" sexual function and mating, a pattern unknown among wild foxes, but typical in domesticated species. In addition, morphological characters appeared that are not found in wild animals, but which are quite common in some breeds of dogs. These include drooping ears, turned-up tails, and perhaps most striking of all, piebald color patterns of black and white, not unlike those found in pinto horses and Holstein dairy cows. All of these traits emerged from a breeding program in which the exclusive focus of selection was behavior, not morphology and not reproductive physiology. Ask for friendly foxes, and you get a whole package of surprises. Although the mechanism responsible for linking these characteristics is not understood, it seems clear that selective pressures operating on one trait may result in wholesale reorganization of developmental programs with unexpected and far-reaching consequences for features unrelated to the selected character.

A strong appeal of this observation is that it reaffirms the concept of organisms as complex, integrated units, rather than a loose assembly of independently selected parts. Thinking of animals in this way helps biologists to understand a wide range of otherwise perplexing anatomical features. For example, breeders of large dogs, such as Newfoundlands and St. Bernards, have long been aware of a tendency for these animals to develop a sixth toe on the front limbs. In most breeds, and in wolves and coyotes, the front limb has five digits. Despite the fact that this characteristic is regarded by breeders as undesirable and has been subjected to hundreds of generations of negative selection, six-toed dogs keep reappearing in the population. In contrast to this situation, small breeds such as miniature poodles and Pekingese often lose the first digit, leaving a four-toed forelimb.

Although functional explanations have been suggested for the sixth toe in large dogs (e.g., it helps St. Bernards when they walk through deep snow, sort of like a built-in snowshoe), they have a hollow ring given the many years of breeding against the persistent pinky. An alternative explanation is that the number of toes a dog has is directly related to body size. Big dogs have six toes simply because patterns of development during limb formation are affected by the number of cells comprising the embryonic foot (i.e., the so-called "limb bud"). The bigger the dog, the greater the number the cells in the limb bud, and the more likely it is that an extra toe will form. Small dogs with few cells in the embryonic limb bud have almost no chance of having a sixth digit and in fact may have only enough cells to generate four toes. The message here is the same as that for piebald foxes: When breeders of St. Bernards select for size, they are indirectly selecting for an extra digit. Ask for a big dog, and you get extra toes.

The two examples given in this chapter are actually variations on a much broader theme—they are but one of many hypothesized mechanisms that account for the evolution of nonfunctional traits. Your assignment is to explore more fully the dimensions of this issue. What other ways have been suggested to account for traits that have no apparent function in wild animals? In developing this essay, you will want to pay special attention to concepts such as the "founder effect" and "genetic drift."

## GETTING STARTED

You should start this essay by reviewing the references listed on the next page. They will give you additional information on the mechanisms by which nonfunctional traits may evolve. The background material in this chapter focuses on only two of these mechanisms, and many others have been hypothesized. Make a list of the mechanisms and jot down a few examples. Piebald foxes and six-toed dogs will give you a couple examples, but many others are available in the literature. Your essay should begin with a clear statement of these mechanisms, along with the relevant examples. After you have discussed these, turn your attention to the question of how likely they are to actually contribute significantly to the evolutionary process. For example, both mechanisms discussed in the background material deal with the process of artificial selection and domestication. Do you think the principles emerging from these studies apply to wild animals in nature? What kinds of circumstances need to prevail in order for them to work? Develop a hypothetical scenario that represents your understanding of how the mechanism might actually operate in a natural population. If you think that the "message" of the stories about the piebald foxes and six-toed dogs has no relevance to evolution in natural populations, explain why.

Finally, in answering these questions, keep in mind the importance of truly testable hypotheses to the scientific process. Can all the hypotheses you have discussed be falsified? How would you go about testing, for example, the veracity of the scenario you developed in response to the questions at the end of the preceding chapter?

## REFERENCES

Alberch, P. 1985. Developmental constraints: Why St. Bernards often have an extra digit and poodles never do. *American Naturalist* 126:430–433.

Belyaev, D. K. 1979. Destabilizing selection as a factor in domestication. *Journal of Heredity* 70:301–308.

Gould, S. J., and R. C. Lewontin. 1979. The spandrels of San Marco and the Panglossian paradigm: A critique of the adaptationist programme. *Proceedings of the Royal Society of London* B 205:581–598.

# *Thermodynamics, Tomatoes, and You* 6

Perhaps nowhere in nature is the magic and mystery of life revealed more beautifully than in the sparkle of a thousand fireflies dancing across the night sky on a warm summer evening. So remarkable is the glittering show produced by these tiny insects that it is tempting to imagine that they live by rules and principles fundamentally different from those dictating the comparatively drab lives of humans and all the other nonluminescent organisms. In fact, the basic principles are exactly the same, and at their core are those pertaining to the constant transformation of energy required to sustain life.

Simply put, one of the definitive characteristics of living things is that they are able to transform energy from one form to another. They acquire it in various forms and they spend it in different ways, but all organisms are unified by the way in which their energy transformations are governed by the same fundamental laws of thermodynamics. The sum of all these internal energy transformations is called metabolism, sometimes poetically described as the "fire of life." For most organisms, the fire of life is nothing more than an allegory for an internal process that is not easily observed from the outside. But in the case of the firefly, the notion of the fire of life is taken a step further and is readily and beautifully manifest as bright flashes of energy that communicate information to potential mates for reproduction. The light is produced in the abdomen when a special compound called luciferin is converted from one form to another. When the so-called "excited" form of luciferin spontaneously decays to a low energy state, particles of light called photons are emitted. The energy that drives the conversion of luciferin is obtained from the universal cellular energy carrier ATP, which, in turn, is also produced by a chain of energy transformations. Ultimately, the fuel for this complex sequence of biochemical events is the food that the insect has eaten.

For humans, the manifestations of metabolism may be less spectacular, but they are in no way less complex. The energy in that spoonful of sugar you put into your coffee this morning will, through myriad metabolic pathways, eventually be transformed into heat, sound, motion, ideas, and a thousand other expressions of energy that are part of our daily lives. For the average person, the heat generated by the fire of life is roughly equivalent to the energy emitted from a 100-watt light bulb. And while it is true that variation among individuals in metabolism may result in some "bulbs" being dimmer than others, it is not the case that this difference in luminosity stems from intellectual firepower. In fact, the thinking process of humans is, energetically, a fairly "cheap" endeavor; it has been estimated by one physiologist that all the energy required for one hour of intense concentration can be obtained in one half of a salted peanut!

Your assignment is to write an essay that compares the energy transformations that occur in animals with those occurring in plants. The animal in this case is you and the transformations are those associated with any activity you chose. Studying for an exam, shooting a basketball, talking to a friend, writing an essay, singing a son—any of these, or any other activity that interests you, is fair game for this assignment. Your analysis should begin with a description of where the energy comes from, how and where it is transformed, and the consequences and effects of the various transformations. As you write your essay, be sure to refer, where appropriate, to the specific form of energy (e.g., chemical energy, heat energy) and to the state of the energy (potential versus kinetic). Finally, consider the implications of the first and second laws of thermodynamics for your energy flow.

When you have finished with the analysis of the energetics of your activity, consider energy flow through a plant. You can select any plant you wish—a vegetable growing in a garden, a tree in a forest, grass in your lawn—any of these will do. Again, describe where the energy comes from, how and where it is transformed, and the consequences and effects of these transformations. The final segment of your essay will consist of a short comparison of the basic differences that exist between you and the tomatoes in your garden in the way energy is acquired and used.

## GETTING STARTED

The first thing you should do is become familiar with the terminology and basic concepts of energy flow in living organisms. Refer to a good textbook, and start reading about energy in the cell biology chapter. Once you have a good understanding of the basic concepts, you will be ready to apply them to the specific case you have chosen.

Next, reflect for a moment on your selected activity. Make a flow chart that describes the major pathway of energy flow. Ask yourself what steps are involved in the activity. Are muscles used? Nerve cells? What is the energetic result of the activity—sound, motion, ideas, etc.? Finally, after you have traced the flow of energy through the major steps, note the points at which it is transformed, how this happens, and where in the body it occurs.

Use the same basic approach to evaluating energy flow through your plant. Set up a flow diagram that shows the flux of energy, how it is transformed, and where that happens.

As a final point of discussion, you may wish to deal in more detail with the second law of thermodynamics (i.e., with every transformation, some energy is lost to the organism in that it cannot be used to do useful work). How efficient are organisms at transforming energy in your selected activity, and how does that efficiency compare with that of machines performing the same function? For example, how does the energetic efficiency of a walking human compare to that of a car, or a human on a bicycle? Or, another example, how does the efficiency of a human speaker compare to that of the stereo system in your dorm room? The answers to these questions are more complicated and will require you to go beyond the material available in an introductory biology textbook. Your instructor can help you in developing your ideas about these considerations.

## REFERENCES

A good introductory biology textbook will have most of the information you need for this essay. For information on efficiency of energy transformations, you may need to consult a biochemistry textbook and/or a textbook on comparative or environmental physiology.

# *Sociobiology and the Evolution of Culture* 7

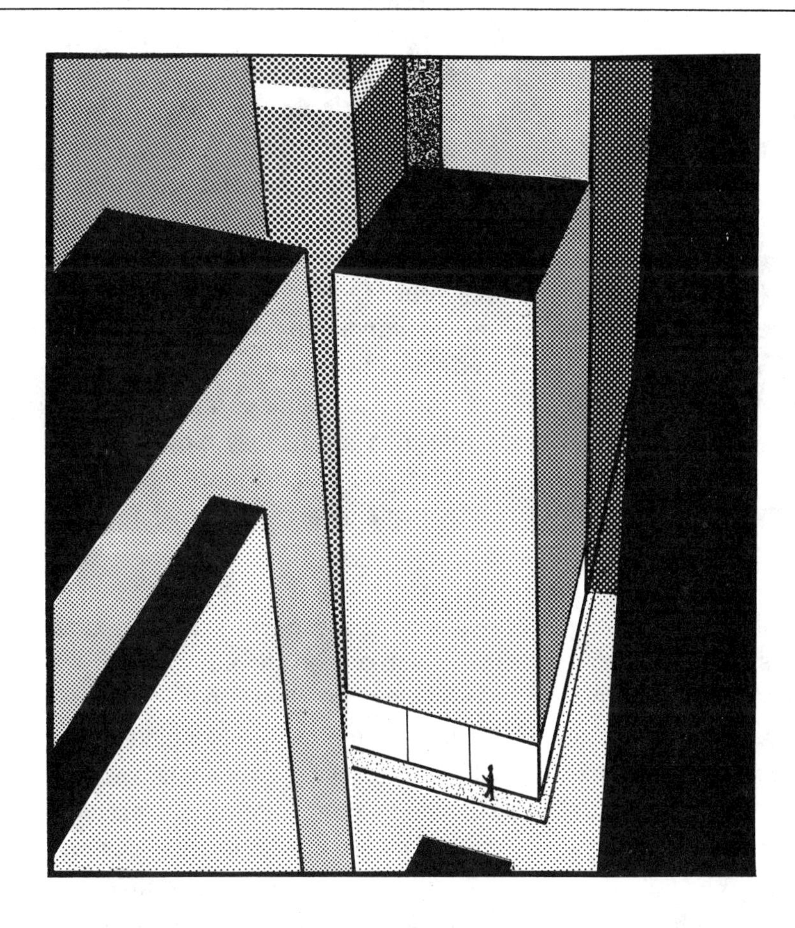

Imagine the following scene. You are alone. It's late, and you're walking home at night, down a dark, quiet street in a sparsely populated neighborhood. Few cars pass. Yet suddenly, as you approach a street light, you see the figure of someone standing next to the pole. Because the figure's back is to you, and because the person is wearing a full-length trench coat, you can't tell who it is, or even if the person is male or female. You stop. You consider your options, but decide to keep walking. As you approach closer and closer, the person next to the pole finally hears the soft echo of your footsteps. The figure turns and stares. You see the face in the light—a woman's face.

If you're like most people, you feel relieved. Without actually thinking about it, you doubt that she will be aggressive or try to harm you. In our society, women commit far fewer aggressive crimes than men. Unconsciously, these considerations guide your actions, and you proceed toward the figure. You may give her a wide berth, perhaps even averting your eyes, although you still keep track of her out of the corner of your eye. But you continue on your way.

Now, for contrast, imagine what your reaction would be if, when the figure turns and stares intently, it's a man standing next to the street light pole. In most instances, you will probably feel more threatened. You fear a confrontation, maybe a violent one. This cautious and fearful perception emerges from the knowledge that men are usually more aggressive than women, they attack more often, and they commit more violent crimes. The anxiety provoked by the encounter might be enough to make you completely reroute your journey home.

The difference in the typical reactions to these two scenarios is not hard to understand from the perspective of self-preservation; people expect and fear a heightened capacity for aggression in men. What is not so clear is why men are so much more aggressive and how that proclivity originated. A relatively new field of study in biology, called sociobiology, offers controversial answers to those questions. Male aggression, and its extreme expression—criminal behavior—is a reflection of millions of years of human evolution in which natural selection has favored so-called "aggressive dominance" social systems, with males almost invariably dominant over females. Not surprisingly, claims that complex cultural phenomena such as crime and gender roles can be understood in terms of genetic makeup have angered and alarmed a veritable army of critics. In some ways, these controversial aspects of sociobiology are regrettable because they tend to detract from the monumental impact it has had on the study of animal behavior. This chapter will introduce you to the major themes of sociobiology and how they can be used to study animal social systems. Your writing assignment is to explore the legitimacy of applying these themes to human behavior and the evolution of culture.

E. O. Wilson, one of the founders of the field, defined sociobiology simply as "the study of the biological basis of social behavior." Animal behavior, according to the principles of sociobiology, is viewed and analyzed in the context of how successfully it promotes the propagation of genes. The focus in these analyses, then, is the gene, not the organism. As Wilson wrote, "In a Darwinian sense, the organism does not live for itself. Its primary function is not even to produce other organisms; it reproduces genes and it serves as their temporary carrier." This concept depicts an animal almost as if it were a machine, the sole function of which is to protect and propagate the genes that ride around inside it.

In many ways, sociobiology has been brilliantly successful as a guiding light for studying the functional significance and evolutionary origin of animal social systems. New insights about old questions that have long puzzled behavioral scientists have resulted from this "gene's-eye view" of evolution and behavior. One important example involves "altruistic" behavior, in which an animal behaves in a way that diminishes its own reproductive success while increasing that of another individual. On the surface, it would appear that natural selection would hammer that kind of behavior and the unfortunate genes responsible for producing it into oblivion. After all, wouldn't altruistic individuals consistently breed less and produce fewer offspring than their less socially compliant competitors? And yet, nature seems to be overflowing with examples of altruistic behavior, from the suicidal sting of a honeybee to the cooperative breeding systems in many species of birds in which some individuals serve as "helpers" to raise offspring that are not their own.

The solution to this puzzle arrived with the realization that related individuals share common genes, and the more closely related they are, the greater their shared heredity. Thus, an individual can increase its genetic contribution to the next generation by helping relatives, with whom it shares genes, to reproduce. Natural selection will favor altruistic behavior (and the corresponding genes) if the behavior causes an increase in the survival of genetically related individuals. W. D. Hamilton coined the term "kin selection" to describe this process, and the concept has proven to be extremely powerful in explaining the enormous variation in complexity and design of social interactions found among species of animals.

If the architects of sociobiolgy had stopped there, the world may not have been a better place, but it would have been spared a fierce and ugly debate, and you would have been spared an essay assignment. But stop they did not, and the sciences of sociology, psychology, anthropology, and scores of related disciplines have never been the same. The principles of sociobiology were extended to human social behavior and culture. A tempestuous maelstrom of contention erupted that continues today, although the unrestrained passions and ideological fervor seem to have abated from the volatile levels reached in the late 1970s and early 1980s.

On topics from gender roles to jurisprudence, sociobiologists have made claims that variously (and occasionally simultaneously!) annoy, anger, fascinate, frustrate, challenge, and insult. For example, prominent sociobiologists have argued that gender roles in modern society have a genetic origin, dating back to hunter-gatherer societies in which men hunt and women stay at home. Hence, regardless of educational and legislative initiatives, men will continue to dominate political life, science, and business. To any woman engaged in the struggle for economic and political equality with men, these claims go far beyond annoying. They constitute an insidious and demoralizing justification of a status quo that perpetuates special privilege based entirely and arbitrarily on gender.

This incendiary beginning notwithstanding, the application of evolutionary principles to human societies has also generated fascinating new perspectives on subjects as divergent as child abuse, inheritance laws, incest taboos, and psychoanalysis. Many of the hypotheses that emerge from these studies are being tested by analyzing patterns in sociological data.

Your assignment is to write an essay that presents your reaction to these issues. In what ways do sociobiological principles appear to explain human social interactions? Where do they fail? In writing your essay, focus on two examples from the literature, one that you think is legitimate (i.e., you are convinced by the study that sociobiology provides a reasonable explanation for a social or cultural phenomenon) and one that you think is not legitimate. In defending your view of these studies, be sure to address the criticisms that have been leveled against human applications of sociobiological principles. Why do you think these criticisms do not apply to the study you regard as legitimate?

## GETTING STARTED

It is critical that you begin this assignment by reading some general articles about human sociobiology. Try, if possible, to read with an impartial view. Avoid jumping to a quick conclusion.

Once you feel confident that you understand the essence of the arguments, try to identify two studies that meet the requirements outlined above—one that you find convincing, and one that you find not so. As you develop your thoughts about these studies, try to evaluate their broader consequences. What bearing might they have for our judicial system? Our family structure? The role of women and men in business, science, or politics? One of the more virulent charges that have been filed against human sociobiology is that of racism. Do you perceive racist thinking in the studies you have chosen? Could the data or ideas presented therein be used to advance racist philosophy?

You may discover, after thoughtful consideration, that you cannot find any but the most trivial of roles for evolution in the social and cultural affairs of humans. That's fine, but you should support your position with carefully reasoned arguments, critical thinking, and appropriate references. Conversely, you may decide that everything you read about human sociobiology makes sense. Again, no problem, as long as you can dispel effectively the counterclaims rendered by critics.

## REFERENCES

The field of sociobiology has a large and extensive literature. The sources listed below will introduce you to the field, but your instructor and a good introductory biology textbook may have additional suggestions. Because human sociobiology is such a dynamic and rapidly developing discipline, broadly circulated journals, such as *Science, Nature,* and *Science News,* have articles or news reports on this subject almost on a weekly basis. These reports are a great way to pick up information about the results of the most recent studies.

Beckstrom, J. 1989. *Evolutionary Jurisprudence: Prospects and Limitations on the Use of Modern Darwinism Throughout the Legal Process.* Urbana: University of Illinois Press.

Bower, B. 1991. Oedipus wrecked. *Science News* 140:248–250.

Boyd, R., and P. J. Richerson. 1985. *Culture and the Evolutionary Process.* Chicago: University of Chicago Press.

Dawkins, R. 1976. *The Selfish Gene.* Oxford: Oxford University Press.

Fisher, A. 1991. A new synthesis comes of age. *Mosaic* 22:10–17.

Lewontin, R. C., S. Rose, and L. J. Kamin. 1984. *Not in Our Genes: Biology, Ideology, and Human Nature.* New York: Pantheon.

Lumsden, C., and E. O. Wilson. 1981. *Genes, Mind, and Culture.* Cambridge: Harvard University Press.

Simon, H. A. 1990. A mechanism for social selection and successful altruism. *Science* 250:1665–1668.

Wilson, E. O. 1975. *Sociobiology: The New Synthesis.* Cambridge: Belknap Press.

# *Unwelcome Newcomers: The Ecology of Introduced Species* 8

In 1986 a ship taking on cargo in the St. Clair River dumped its ballast water. That in itself was not an unusual occurrence. Ballast water is used to stabilize a ship and make it seaworthy. Most ships routinely add or discharge ballast water in order to accommodate loading or unloading. At that time, there were no regulations in the United States governing the intake or discharge of ballast water.

On that particular day, however, it's likely the discharging ship had taken on the ballast water in freshwater ports around Europe. Over the period of a day or two, the ship flushed thousands of gallons of water into one of our navigable rivers. Along with the water, it also introduced billions of foreign organisms, most of which died immediately or in a very short time thereafter.

One species persisted, however. The species is, at first glance, a fairly innocuous invader. In its adult stage it is nothing more than an inch-and-a-half-long mollusk. To make it even less formidable, the St. Claire River and the Great Lakes water system, in general, already had numerous species of mollusks. For nearly two years the zebra mussels, as the invaders are known, were so obscure that they went undetected. That was soon to change in a very big way. In 1988 a curious biology student plucked one of the striped shells out of Lake St. Clair. Soon thereafter it became evident that the zebra mussels, in less than two full years, had become an established invader of the Lake St. Clair ecosystem. The next stage of range expansion was a simple matter of following the river downstream. As scientists predicted, zebra mussels had been swept down the Detroit River and had begun to colonize Lake Erie as well.

Again, this didn't seem to be a major concern, until the full impact of the species became evident. The zebra mussel reproduced in tremendous profusion. A single female can pump 40,000 eggs into the water column. The resulting population growth can be spectacular. In 1988 biologists at the Detroit Edison plant on western Lake Erie counted 200 mussels per square meter on the intake screen. The following year the number had increased to 700,000.

Ecological considerations aside, the financial costs of the zebra mussel invasion are staggering. By reproducing so rapidly and overwhelming the aquatic communities where they are found, the zebra mussels have created a host of problems for cities and towns on Lake Erie. Clogged water pipes, reducing water flow in some areas to 25 percent of normal rates, have forced the shutdown of hospitals, emergency stations, and some schools. The municipal costs of refitting pipes, restructuring water in-takes, sewers, and so on, are calculated in billions of dollars. Moreover, because the zebra mussel filters huge amounts of water, sucking out most of the phytoplankton, the lower food chain may be depleted, with the resulting effects reverberating up the food chain. Walleye, a popular game fish, are threatened by the loss of spawning beds to the enormous mass of zebra mussels encrusted there. Walleye fishing in Lake Erie alone is a 900 million dollar a year business.

Ecologically it is almost impossible to predict what the full effects of the zebra mussel invasion might be. Obviously native species of clams are in jeopardy. Zebra mussels literally cover the native clams, thereby ensuring first claim to nutrients in the water column. By putting such extraordinary competitive pressure on native species, the unusual biological diversity existing in the Great Lakes mollusk community—including 18 different species—is being rapidly undermined. Several biologists hypothesize that the zebra mussel may soon be the only mollusk in all of Lake Erie.

Ironically the water clarity has improved in Lake Erie. Because of their voracious appetites, the zebra mussels have doubled clarity levels in just a short time. This same appetite, however, may prove its own downfall. In a real sense the zebra mussels are eating themselves out of house and home. In fact, many biologists insist that the zebra mussels will continue to grow until they reach an unsupportable population level. At that point, the speculation goes, the zebra mussels will experience a population crash. The proponents of this scenario believe the question is not how high the population will rise in the early stages of invasion, but where it will end up after the crash.

The zebra mussel, whatever the final outcome of the invasion, is certainly not the first case of a species successfully colonizing new habitats in the United States in dramatic fashion. Some notable invaders have been the sea lamprey in the 1830s, the alewife in the 1870s, and the tube-nosed goby in 1990. The most widely known example, however, may be the common starling. A hundred years ago a bird fancier introduced 50 pairs of starlings to New York's Central Park, and today the starling is the most populous bird in the United States. The starling serves to remind us of how quickly, and efficiently, some species adapt to a new environment.

In all these cases, the introduced species have flourished spectacularly in their new habitats. In evaluating the history and consequences of introductions, it is clear that some species of plants and animals are especially well suited for invading a new biological community and, conversely, that some communities are especially vulnerable to invasion. Your assignment in this chapter is to write an essay in which you provide specific examples of invasion by introduced species, and then assess the circumstances and characteristics associated with a successful invasion. In addition, you should also consider the question of the ecological consequences of an introduction. How predictable are such consequences? Given what we know about the fate and impact of introduced species, under what circumstances can you justify the deliberate introduction of species into new habitats?

## GETTING STARTED

As a starting point you should try to identify the characteristics of a successful invader. What aspects of reproduction, dispersal, gene pool, competitive ability, or physiological tolerance predispose an organism for success when it is introduced to a new environment? Do the characteristics you have identified apply to both plants and animals? To insects as well as vertebrates? In answering these questions, you should also try to gather information by considering introductions that were not successful. Are there unifying features found among those species that failed to thrive when presented with the opportunity to colonize new habitats? Based on your judgment of the literature on introduced species, what hope is there of some day developing a comprehensive model that will be able to predict the outcome of an invasion?

Your essay should also consider the characteristics of biological communities that appear especially resistant or especially vulnerable to invasion by new species. Does the location of the habitat make it more or less vulnerable to invasion? For example, is Florida at greater risk in the introduction of foreign species than New England? Why? What about the Hawaiian islands in contrast to islands in the Caribbean? How can you account for the differences in vulnerability?

Finally, what happens to community structure after an introduction occurs? How often are native species driven to extinction? How often are the results completely unpredictable? Given this level of predictability, provide an evaluation of the wisdom of deliberate introductions. Can you find examples where the intentional introduction had the desired effect? How about examples of a catastrophe stemming from well-meaning ventures of this type?

## REFERENCES

Drake, A. J. (ed.). 1989. *Biological Invasions: A Global Perspective*. New York: Wiley.

Griffith, B., J. M. Scott, J. W. Carpenter, and C. Reed. 1989. Translocation as a species conservation tool: Status and strategy. *Science* 245:477–480.

Miller, J. A. (ed.). 1988. Hawaii's unique biology. *Bioscience* 38:232–282.

Palca, J. 1990. Libya gets unwelcome visitor from the west. *Science* 249:117–118.

Roberts, L. 1990. Zebra mussel invasion threatens U.S. waters. *Science* 249:1370–1372.

Spencer, C. N., B. R. McClelland, and J. A. Stanford. 1991. Shrimp stocking, salmon collapse, and eagle displacement. *Bioscience* 41:14–21.

Vitousek, P. M., L. L. Loope, and C. P. Stone. 1987. Introduced species in Hawaii: Biological effects and opportunities for ecological research. *Trends in Ecology and Evolution* 2:224–227.

# *Human Rights, Animal Rights, and Species Rights* 9

Most people understand and accept that humans have a responsibility to treat animals in an ethical manner. At the same time, few question the legitimacy of the right people have to use and develop the natural resources available to them. And further, although you might quibble with terminology, few people would argue with the view that species have a right to exist, and that efforts should be made to ensure their survival. What isn't so clear is how to resolve the competing interests found in the large gray area where these rights collide. Your assignment in this essay is to study the following examples and try to establish reasonable criteria on which to base the difficult decisions that arise when these legitimate rights are in conflict.

1. In New Hampshire the moose population has increased recently in dramatic fashion. Twenty years ago biologists feared for the moose's survival in the northeastern portion of the United States. Although the population remained fairly strong in Maine, the rest of its range—New Hampshire and Vermont—had succumbed to development and intense human population pressure.

Recently, however, things have changed. The moose population has rebounded in sufficient numbers to allow the Department of Fish and Game to establish a limited hunting season. Each year the Department of Fish and Game holds a raffle to determine who can "harvest" the moose. Since moose can weigh up to 900 pounds, and the meat is highly prized, the hunting lottery is hotly contested. Given recent economic times, however, especially in the beleaguered northeast, the moose meat is a welcome addition to any family's food budget.

What seems to have been a successful campaign to reinstate the moose to part of its ancient range has not been without its side effects. Animal rights activists are calling for a cessation of the moose hunt, calling it cruel and unnecessary. They find a particularly dark symbolism in the idea of a lottery. To end an animal's life, they claim, by spinning a barrel of tickets and then drawing a select few, is the height of human indifference to other species.

That's where the issue stood for several years. The hunt went on; the animal rights groups protested in the state's capitol. Then another factor entered the equation.

During autumn and spring, moose began crossing the highways in great numbers. An adult bull moose may chase a yearling out of his territory, so the yearling, his own senses quickened by the possibility of mating with a female, stumbles through the woods in a state of confusion. A road, and its attendant fencing, is nothing at all to a healthy moose. Since moose have no natural enemies, and since they see well at night, they often wander onto the roadways where collisions with motorists are inevitable. On a stretch of Interstate 93, a four-lane road running north and south through New

Hampshire, 170 collisions have occurred in the last three years. The collisions have also resulted in approximately 20 human fatalities.

As a result, the animal rights activists have had steep going when arguing against the hunting lottery. The sentiment seems to be that animals are okay as long as they remain in the woods where they belong, but when they begin to interfere with our natural enjoyment of the roads or countryside, then they are nothing but nuisances. To add a further complication, some biologists now claim that the moose are running off deer herds, which are, by far, a more valuable commodity than the moose. The economy of northern New Hampshire depends on the hunters for a shot in the arm each fall. Without the hunters—who come in greater numbers for deer than moose—the economy stands to suffer. What are the humans' rights? The deer's rights? The moose's rights? What should be done?

2. In July of 1987 a snorkeler in Wakulla Springs, Florida, was attacked by an alligator. The alligator rolled the man in a death shake, then carried the body under water and stored it beneath a riverbank. It took several days for additional divers to find the snorkeler's body. The alligator was never captured.

But other alligators were. In a knee-jerk reaction, people around the state of Florida began killing alligators indiscriminately. It was, as some biologists termed it, the "Jaws" phenomenon. One shark is bad from a human perspective, so all sharks are bad. One alligator has killed a man, so all alligators must be slaughtered.

Statistically it's surprising more attacks don't take place. Between 1970 and 1980 Florida's human population jumped from 6,792,417 to 9,739,992—a 43.4 percent increase. Eight hundred people move into the state every day. As a result, nearly 12 million acres of wetlands have been lost to wildlife so far. Some 800 acres are drained or filled each year.

With all the development and population pressure, it's likely humans and alligators will come into conflict more often in the near future. While the human population of Florida has undergone a gigantic increase, so, too, has the population of alligators. As recently as 1969 the federal government placed alligators on the endangered species list. By 1977, however, the population had rebounded so well that they were taken off the list. In the last two or three years, regulated hunting has been allowed in most of Florida, Louisiana, and Texas, the states that comprise the alligator's chief range.

Something has to give. City planners are extremely nervous about the lagoons and canals that many Florida homes have running through their land. Children will naturally be attracted to the canals in the warm climate, and invariably alligators will be there waiting. Although alligators are not particularly territorial, they still represent a considerable danger. Recent television and newspaper accounts reported the story of a little boy who was the object of a tug-of-war between his frantic mother and a large alligator. The mother barely won, but the image will remain in many people's minds forever. What are the alligators' rights? What are the human's rights? What should be done?

3. In 1972 the U. S. Navy initiated a program to control the population of goats on the Channel Islands off the coast of California. It's unlikely the navy had any idea what it was getting into. As soon as they announced the program, animal rights activists began to protest what they perceived as the wanton destruction of animals. The navy responded by saying it was only trying to help. The idea of shooting the goats had actually sprung from biologists who were concerned that the goats—as goats have done to islands throughout the world—were decimating the natural ecology of the Channel Islands. The goats, they pointed out, were introduced to the islands. They were not indigenous. Since no natural predator existed to keep their number in balance, they were thriving while other species on the islands declined as their habitats were destroyed by the voracious goats. In fact, if the goat population continued to expand unchecked, the consequence was likely to be extinction for dozens of endemic island plants and animals.

Never mind, the animal activists cried, you still can't kill the goats. The debate rapidly developed into a heated, often emotional exchange of views. The animal rights activists claimed that the goats, however destructive they might be, were guilty of nothing but acting like goats. To kill them, the activists claimed, was willful and interfering. Didn't human interference put the goats there in the first place? What right did we then have to go back and try to erase the mistake by slaughtering thousands of innocent animals?

The biologists, on the other hand, felt the goats posed such a threat that they needed to be eradicated. To cut back their numbers was futile; they would quickly repopulate. In the meantime, the biologists argued, the animal activists were showing crass insensitivity to the rights of the indigenous species. What are the goats' rights? The animal activists' rights? The indigenous species' rights?

## GETTING STARTED

These three examples are meant to demonstrate some of the sticky issues now being debated concerning species rights. We make decisions about animal rights everyday by the food we eat (beef, chicken, pork) and by the clothes and ornaments we wear (leather, fur, pearls, animal-tested cosmetics). Occasionally, however, certain cases arise that force us to look at our position on these issues. Perhaps one of these cases, after further research and thought, will crystallize some of your own attitudes.

The best way to approach this essay might be to come up with your own personal guidelines concerning animal rights. Do you feel each case must be decided on its own merits? Or do you feel in these times of runaway population growth that we must come to personal ethical decisions about our position regarding animals? Are there any lines to draw?

Think about these matters carefully. You may want to begin with an examination of your daily habits. Do you eat meat? Do you wear leather, wool, animal-tested cosmetics? Would you wear an exquisite fur coat if someone gave it to you? Where do your rights end and an animal's rights begin?

## REFERENCES

Anonymous. 1990. Whose hills? *Economists* 315:69–70.

Praetzel, Anne-Marie. 1990. Island allure. *National Parks* 64:35–37.

Prindle, K. S. 1988. Alligators and people: When two populations collide. *The Animals' Agenda,* January/February, pp. 32–34.

Sleeper, B. 1989. Who gives a hoot about the spotted owl? *Animals* 122:25–27.

# *Unit III Biology, Technology, and Society*

Advances in biological research are proceeding at lightning speed. It used to be said of the biological sciences that the difference between basic and applied research was about 20 years. In today's world, the difference is more like two weeks. Fundamental discoveries in basic research that are made without reference to human applications can often be transformed, almost overnight, into an applied concept that has direct implications for the human condition. This unit is about those concepts and the impact they have on human lives. Separate chapters describe the biological principles that have opened new horizons in fields as widely disparate as genetic engineering, human sport performance, drug abuse, human reproduction, and pesticide application. However, in many of these cases, the proffered "brave new world" is not a vision of unmixed blessings. Each comes with its own positives and negatives, its obvious benefits, and its not-so-obvious drawbacks.

The use of pesticides may increase the productivity of a farmer's fields or the beauty of a suburban lawn; but as you will discover, these come with an environmental pricetag. Genetically engineered organisms have enormous potential to enhance agricultural capacities for crop production, but can we safeguard these techniques to prevent the catastrophic release of dangerous new organisms into the environment? Advances in exercise physiology have led to greater understanding of what happens when a body exercises, but these in turn have created new opportunities for cheating through artificial enhancement of human performance, compromising both the health and the integrity of the athlete. And we are rapidly approaching the point in human reproductive physiology where it will be possible to select the gender of your child. What impact is this likely to have on future generations? Lastly, in the area of recreational drug use, have the recent advances in understanding how drugs work in the cells of the brain created a new and even more dangerous generation of "designer drugs"?

These issues will be the focus of your writing exercise in this unit. More than the first two units, you will need to develop your understanding of the principles through background reading. As you read the material, try to keep in mind the many dimensions and societal implications of the biological principles under consideration. Where appropriate, try to come to a conclusion regarding the overall social impact of the technology.

Then, go write!

# 10 Designer Genes

Tomatoes have come a long way, figuratively and literally, since the early Spanish explorers first hauled them from the New World back to Spain. In fact, for the next three centuries, they were regarded with great suspicion by most Europeans and New World colonists, due primarily to the bad reputation of a couple of highly poisonous close relatives, deadly nightshade and mandrake. As recently as the turn of this century, educators and scientists, prominent among them George Washington Carver, performed dramatic public demonstrations of tomatoes being eaten in an attempt to promote their cultivation and consumption. It wasn't until the nineteenth century that "love apples," as the French called them, really began the long climb toward respectability on the list of plants cultivated for food.

From that tentative beginning, tomatoes have become one of the most popular food plants in the world. Barbecued hamburgers and pizza, among other features of contemporary life, have not been the same since the cultivation of tomatoes began in earnest. The popularity of this garden "vegetable" (technically a fruit, popular description notwithstanding) has made it a prime candidate for plant breeding programs, resulting in the diverse array of colors, shapes, sizes, and names found in any garden shop or fruit stand during the summer. From "Super Girls" to "Big Boys" to "Red Cherries" to "Golden Boys" to "Early Girls" and all the way to "Better Boys," the tomato's intrinsic genetic diversity in fruit size, color, and timing of maturity has created a veritable dictionary of options for the home gardener. There are, however, limits to the extent of genetic diversity in all organisms, and tomatoes are no exception. Genetic engineers have begun to explore the boundaries of genetic limitations in tomatoes, and the initial results of their research are highly encouraging. In a single generation, plant geneticists have been able to introduce traits in tomato plants that might have taken a thousand generations to develop through standard breeding practices.

Such results are becoming increasingly common—basic research in the field of biotechnology is progressing on virtually all fronts at a breathtaking pace. Hardly a week passes without news of some new advance in areas from medicine to agriculture. In fact, agricultural applications have received a great deal of attention because of the potential for dramatic improvement in crop productivity. Such applications have also received attention because of the perceived environmental threat posed by this technology. Your assignment in this chapter is to explore both sides of the debate over the synthesis and environmental release of genetically engineered organisms, and then come to your own conclusions regarding the impact and advisability of this new technology.

Genetic engineers have developed a number of novel ways for increasing the productivity of cultivated plants. One such approach involves the insertion of toxin-producing genes from a bacterium, *Bacillus thuringiensis,* into tomato plants. The gene is "turned on" inside the plant cells, producing a toxin that kills only insects. The result is that the plant acquires immunity from attack by insects, eliminating the need for costly and environmentally questionable applications of chemical pesticides.

Another advance in tomato cultivation is a genetic alteration that extends the length of time before the ripened fruit will rot. Hence, tomatoes can be stored and shipped after they have ripened on the vine, in contrast to the current practice of picking the fruit green, then shipping it. By some estimates, 50 percent of a tomato crop is lost to spoilage due to the difficulties of timing the harvest of the fruit. Much of this wasted food resource can be reclaimed by genetically manipulating the developmental process of fruit maturation.

Yet another genetic engineering advance that can aid agricultural productivity relies on the insertion of a gene for resistance to the widely used herbicide, glyphosate (known best by the tradename Roundup). After the gene is inserted into the plant, the plant acquires resistance to the herbicide, and workers can then "weed" the field by spraying it with herbicide. The weeds will die, but the genetically engineered plants will remain unharmed.

Advances such as these, however, are not universally perceived as an "unmixed blessing." Expressions of caution and concern have been raised by a number of scientific and public organizations regarding the release of genetically engineered organisms. These concerns have created a strict regulatory process to prevent the inadvertent release of harmful genetically engineered organisms. Some have argued, however, that the regulatory process is not nearly strict enough, given the enormous risks that are perceived to be involved in such experiments. Suppose, for example, that a profligate and costly weed species were to acquire the glyphosate resistance discussed above. The ecological problems created by such a "super weed" may verge on the catastrophic.

On the other hand, many scientists now feel that these regulations and the lengthy delays they create are excessive and have produced a bureaucracy that is choking off research in this area. Your assignment is to write an essay in which you assess research on genetically engineered tomatoes, both in terms of the potential benefits and the ecological risks. How do genetic engineers safeguard against environmental catastrophes? In your view, are these safeguards sufficient to allow the experimental release of engineered organisms? Or should research of this type be banned entirely on the basis of the enormous environmental threat it poses.

## Getting Started

You should begin your essay with a discussion of the mechanisms underlying insect pest resistance and taste/texture modification in engineered tomatoes. Following that, discuss the environmental safeguards that must be built into any experiment that involves the release of genetically engineered organisms. How, specifically, are tomatoes modified to ensure those kinds of safeguards? Be sure to include in your analysis an assessment of the risks and the benefits to be derived from the products of these procedures. In your opinion, do the benefits outweigh the risks?

## References

Brunke, K. J., and R. L. Meeusen. 1991. Insect control with genetically engineered crops. *Trends in Biotechnology* 9:197–200.

Cavalieri, L. F. 1991. Scaling-up field testing of modified organisms. *Bioscience* 41:568–574.

Fox, J. L. 1990. Environmentalists carp about planned test. *Biotechnology* 8:286.

Fox, J. L. 1990. Herbicide-resistant plant efforts condemned. *Biotechnology* 8:392.

Fox, J. L. 1990. Toward unified rules on deliberate release. *Biotechnology* 8:499.

Geisow, M. 1991. The proof of the cloning is in the eating. *Trends in Biotechnology* 9:5–7.

Hochberg, M. E., and J. K. Waage. 1991. Control engineering. *Nature* 352:16–17.

Lambert, B., and M. Peferoen. 1992. Insecticidal promise of *Bacillus thuringiensis:* Facts and mysteries about a successful biopesticide. *Bioscience* 42:112–122.

Marx, J. L. 1987. Assessing the risks of microbial release. *Science* 237:1413–1417.

Miller, H. I., R. H. Burris, A. K. Vidaver, and N. A. Wivel. 1990. Risk-based oversight of experiments in the environment. *Science* 250:490–491.

Moffat, A. S. 1991. Research on biological pest control moves ahead. *Science* 252:211–212.

# *Have It Your Way: Ethical and Demographic Consequences of Gender Selection in Human Reproduction* 11

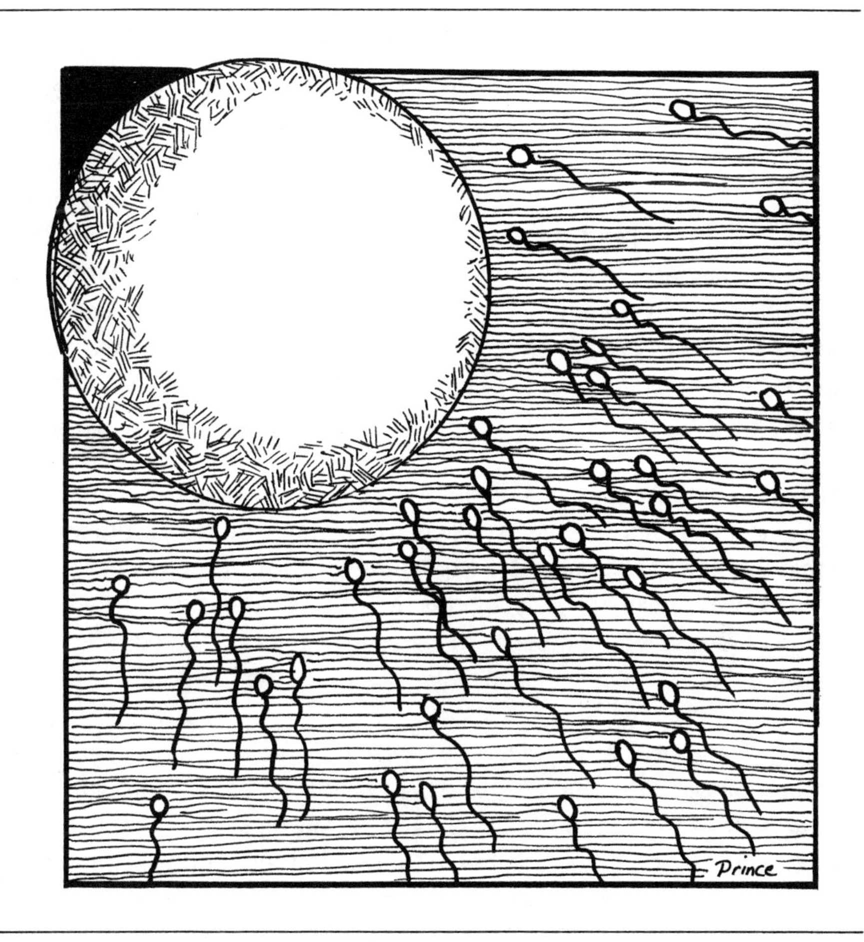


If you could determine the sex of your unborn child ahead of time, would you? Until recently such a question was purely academic, and predicting the sex of an unborn child was an innocent parlor game. One common home diagnostic test consisted of a simple darning needle dangled from a piece of thread alongside a pregnant woman's spine. If the needle spun clockwise (or counterclockwise, depending on who administered the test and, perhaps more importantly, on what the expectant mother hoped to hear) the unborn infant was sure to be a male. Predictive powers of that kind have been associated through folklore with countless other conditions during pregnancy. If a pregnant woman's skin was prone to blemishes, then the child was likely to be male. If the child made the mother more beautiful and glowing, a highly subjective test to be sure, then the child was naturally female. A kicking baby was a male. An even-tempered baby was a female.

Folklore of this type is not restricted to simply predicting the sex of a developing fetus—it also extends to claims regarding gender determination before fertilization. Many factors, including frequency and timing of intercourse, sexual behavior, nutrition, and even that durable old chestnut (usually offered in jest), "something" in the drinking water, have all been implicated as significant determinants of which kind of sperm, male-producing or female-producing, will fertilize the egg. Formal evidence for the efficacy of these folk practices is, of course, virtually nonexistent, although theoretical explanations have been postulated in several widely distributed books.

Whether these folk practices actually work or not will soon be an irrelevant question. Advances in human reproductive technology are occurring at such a rapid rate that many scientists expect that preconception gender selection will become a reality in a very short time. Your assignment in this chapter is to write an essay that addresses the sociological, cultural, and demographic consequences of this advance in biological technology.

To understand how it may soon be possible to select the gender of one's offspring before the egg is fertilized, some background information about sperm is necessary. Gender in humans is determined by the type of sex chromosome contributed by the father. Each human egg contains 23 chromosomes, and one of these is the X-chromosome. Each human sperm also contains 23 chromosomes; one of these is either an X or a Y. If the egg is fertilized by an X-sperm (gynosperm), then the child is female (XX). If the lucky winner of the fertilization sweepstakes is a Y-sperm (androsperm), then the child is male (XY). Androsperm and gynosperm differ in several ways, including shape, size, and total DNA content (due to the relative size the X- and Y-chromosomes). Because androsperm are somewhat smaller and have longer tails, they tend to swim faster than gynosperm. Moreover, the average production ratio of androsperm to gynosperm is roughly 2 to 1. If androsperm can outswim gynosperm, and they outnumber the female-producing sperm by a 2 to 1 margin, then why isn't there a disproportionate number of males in the human population? In fact, androsperm do have a marked advantage. At conception, there may be as many as 160 males for every 100 females. Male embryos, however, are considerably less hardy than female embryos, so at birth the ratio is approximately balanced: 105 males for every 100 females.

Another factor is the acidic environment of the vagina. The advantages of androsperm are undermined by the fact that gynosperm are less vulnerable to the acidic environment of the vagina. The more acidic the environment, the more disadvantaged are the Y-sperm in the great sprint to fertilize the egg. This observation has been invoked in support of several of the anecdotal approaches to gender selection mentioned previously. Nutritional aspects of the diet, for example, may affect a woman's internal acid/base balance and thereby lead to a predisposition for either male or female offspring. In a similar fashion, it has been claimed that the timing of intercourse can influence offspring gender because the acidity of the vaginal environment changes during the menstrual cycle. Although these scenarios seem logical (and that "logic" has sold thousands of books to couples trying to select the gender of their baby!), it nevertheless is the case that solid experimental evidence in support of these claims does not exist.

What does exist in research laboratories all over the world is an enormous array of biochemical methods that can be used to separate and isolate small biological particles on the basis of slight differences in structural or chemical properties. These are the techniques that offer the hope (or the horror, depending on your view) for real advances in gender selection. The idea is that these methods can be used to separate Y-sperm from X-sperm, followed by the use of artificial insemination to obtain the desired gender in the fertilized egg. If females are desired, inseminate using semen enriched with X-sperm. For males, use androsperm-enriched semen. As yet, no single technique has been unequivocally shown to produce the desired semen separation, but this area of research has become extremely active, in part because of the application to humans but also because of the interests of the cattle industry. Cattle breeders have a strong financial incentive to find a good sperm separation technique. For dairy cattle, only females are needed; for beef, only males are desired.

Some day in the not too distant future, couples will have the routine option of selecting either a boy or a girl. The focus of your essay will be to explore some of the many ramifications of this reproductive option. What are the positive consequences of sex choice? The negative? How common do you think gender selection will be? Would you personally choose such an option? How about your friends and acquaintances? How might this new reproductive technology affect the demographic profile of the United States? Of developing countries? It has been suggested, in an argument labeled inflammatory and racist by incensed dissenters, that widespread and readily available gender selection is the perfect panacea for the global overpopulation problem. Do you agree? How might social relations change as a result of this technology? Are women likely to outnumber men, or will it be the other way around? What kind of characteristics do you think would develop in a society with a preponderance of males? Of females?

What about the morality of this situation? Do gender selection techniques violate your sense of morality or your idea of "reproductive freedom"? What agencies—governmental, religious, private, corporate—should support this research? Selecting the gender of an unborn child brings us perilously close to what has been called "nature's will" or the "hand of God." If we begin delving into gender selection, aren't we, in some sense, tinkering with a natural balance achieved by nature through the millennia? And yet everybody knows a couple who, after producing three boys, are desperate to have a little girl. Does it seem fair to withhold information that might make their lives happier and more fulfilled? At a more general level, doesn't a woman have the inherent right to control her own reproduction? And what about the claim of a feminist leader that gender selection is "one of the most stupendously sexist acts" possible, because it judges the worth of a human being first and foremost on the basis of sex.

## Getting Started

If you're like most students, the preceding discussion has left you with a bewildering array of questions and not nearly enough answers. To write this essay effectively, you are going to have to organize your thoughts, focus your attention on a limited subset of the questions posed above, and think critically about the questions at hand. The best way to start is to make two lists with positive aspects of gender selection on one and negative aspects on the other. Try to think of as many as you can. When you identify a "positive," try to imagine how that same feature could be viewed as a "negative." For example, gender selection may lead to the elimination of debilitating sex-linked diseases, such as hemophilia and Huntington's chorea, by preventing the production of children of the sex at risk. But, how close might special breeding programs, adopted on that basis, come to the horrors of state-imposed eugenics?

When you have finished your list, focus your attention on one aspect of this issue. For example, you may wish to concentrate on the question of demographic and corresponding cultural changes that may ensue with readily available gender selection. To get a sense of what these changes might look like, interview a number of your friends and ask them if they would opt for gender selection, and if so, what sex they would choose for their children. Are first-born children more likely to be male or female in the hypothetical families of your friends? What about second-born children? After you have assembled this information, use it to predict how the world might change because of this technology.

Finally, after you have considered the questions and analyzed the answers, present your opinion regarding both the moral and legal aspects of this topic. Should gender selection techniques be outlawed or freely available? If you support legislation prohibiting gender selection techniques, what other "reproductive freedoms" would you curtail? The availability of techniques to improve fertility? The right to an abortion? The number of children a couple may produce? In your mind, do these questions constitute simple variations on the same theme, or are they fundamentally different?

## REFERENCES

Beernink, F. J., and R. J. Ericsson. 1982. Male sex preselection through sperm isolation. *Fertility and Sterility* 38:493–495.

Holmes, H. B., and B. B. Hoskins. 1987. Prenatal and preconception sex choice technologies: A path to femicide? In *Man-Made Women,* G. Corea, et al. (eds.). Bloomington: Indiana University Press.

Fletcher, J. C. 1983. Ethics and public policy: Should sex choice be discouraged? In *Sex Selection of Children,* N. Bennett (ed.). New York: Academic Press.

# *Keys to the Doors of Perception: Neurotransmitters and the Action of Psychoactive Drugs*

In 1982 college-age patients suddenly began to show up at hospitals and clinics throughout California's Silicon Valley with highly unusual neurological disorders. These included tremors, uncontrollable shaking, and a general lack of muscular coordination, in some cases so severe as to render the individual paralyzed and unable to speak. Physicians and neurologists had no trouble recognizing this complex of symptoms; to varying degrees, all the young patients displayed the diagnostic characteristics of Parkinson's disease. What puzzled the doctors was the age of the patients. Parkinson's disease is a condition usually associated with advancing age; it is extremely rare in teenagers and young adults.

The mystery was solved when investigators discovered that all the victims had used the same recreational drug, a synthetic pharmaceutical that had been sloppily prepared by a young clandestine chemist experimenting with the manufacture of "designer drugs." His sloppiness in making the drug led to its contamination with a compound called MPTP. Subsequent studies have demonstrated that MPTP attaches to and then kills specific nerve cells in a special area of the brain. For normal, healthy people, these cells remain functional throughout life. In elderly Parkinson's patients, they are lost progressively with advancing age, leading eventually to a debilitating loss of muscle control. In the tragic and unintentional experiment conducted amid the California drug culture of 1982, the death of these cells in healthy young people plunged them literally overnight into permanent, irreversible Parkinson's syndrome. Ironically, the discovery of MPTP in the ashes of this personal tragedy has ignited a revolution in Parkinson's research. Using MPTP as a tool, new drugs are rapidly being developed that have the potential to improve dramatically the treatment of Parkinson's disease.

The story of MPTP carries with it a number of important messages; some of these are highly personal in nature, others are related to fundamental biological principles. Perhaps most strikingly, it reaffirms the claim that most recreational drug users have no idea of what they are actually ingesting and even less information about how the drug works. Your assignment in this chapter is to write an essay that discusses the molecular and cellular basis of the effect of psychoactive drugs, followed by an evaluation of the long-term consequences of habitual drug use. By the time you finish this assignment, you will know how drugs work and how frightening the consequences of their use can be.

MPTP's destructive effect on specific nerve cells begins with its attachment to special receptors in the plasma membranes of these cells. Under normal circumstances, these receptors serve as the site of attachment for chemical messengers occurring naturally in the brain, the neurotransmitters. All of the familiar functions of the nervous system—perception, cognition, and creative thinking prominent among them—result from neurotransmitter-mediated communication among nerve cells in the brain.

The twentieth-century writer Aldous Huxley once described the human nervous system figuratively as a door with two functions. On the one hand, it serves as the portal through which we observe and comprehend our universe, but on the other, it also functions as a filter, sifting information from an otherwise overwhelming barrage of signals emanating from the world around us. Whatever the poetic merits of that expression of human cognition might be, the doors of perception, as Huxley described them, can in reality be found in the billions of interacting neurons responsible for processing information in the nervous system. And if neurons are the doors, then neurotransmitters are the keys, because it is these chemicals that provide a vehicle for the propagation of information throughout the complex network of nerve cells.

Neurotransmitters elicit their effect when they are released by one neuron, travel across the space between cells, and attach to the receptor site on an adjacent cell. The arrival of the neurotransmitter at the receptor site either "excites" the target neuron or "inhibits" it. If the sum of all the neurotransmitter-mediated input to a target neuron produces a sufficient level of excitation, a nerve impulse, called an action potential, is generated and subsequently propagated to the next adjacent neuron where the cycle of neurotransmitter release is repeated.

Dozens of different neurotransmitters, each with an appropriate receptor, have been identified in the brain, and more are discovered every year. The more that is known about these compounds and how they work, the more fascinating they become. For example, a number of nervous system disorders have been linked to neurotransmitter dysfunction. Severe depression is associated with low levels of the neurotransmitter serotonin, and Parkinson's disease characteristically results in low levels of the neurotransmitter dopamine following the nerve cell death described in the opening paragraphs of this chapter. Too much neurotransmitter creates a different set of problems—schizophrenic symptoms may involve excessive dopamine activity. It has even been suggested that certain types of compulsive behavior, such as excessive dieting or fanatical exercise, may be linked to the release of neurotransmitters called endorphins. Endorphins have been described as the brain's "natural painkiller," and they promote a sense of euphoria and well-being. "Runner's high," for example, is thought to result from endorphin release during intense exercise. They were first isolated and identified only after neurobiologists predicted their existence from the discovery of opiate receptor sites, the points of attachment for heroin, morphine, and their chemical derivatives. If the receptors exist, the reasoning went, so too must a corresponding neurotransmitter.

An understanding of how neurotransmitters work has also led to new breakthroughs in the study of the effects and dangers of psychoactive drugs, and these will be the subject of your writing exercise. Your assignment is to write an essay that discusses how any one of several commonly abused drugs affect the way neurotranmitters function in the nervous system. You may choose to describe the action of drugs such as opium, cocaine, LSD, amphetamines, THC (marijuana), "ecstasy," or any other in your essay. Choose only one drug, but be sure to describe unambiguously and with as much detail as possible exactly how the drug interferes with neurotransmitter function. After you have described the short-term effects of the drug, describe the physiological and behavioral consequences of habitual, long-term use. What actually happens to the brain when it is repeatedly subjected to intoxication with the drug under your consideration? Are there side-effects or other neurological complications that can arise from repeated use of the drug?

## GETTING STARTED

You should start with a general outline of how neurotransmitters work, providing the background material for your discussion of the drug you have chosen. When writing about the effects of the drug, keep in mind that you are trying to knit together two levels of analysis. At one level is the effect of the drug as it is experienced by the user, at the other is the effect of the drug as it is experienced by the nerve cells. Your essay should endeavor to describe both of these, then to integrate them. Can specific cognitive sensations be accounted for by analysis at the level of the cell?

As a point of departure in your essay and to broaden its scope, you may wish to contrast the mechanism of action in your selected drug with the manner by which alcohol elicits its effects. How do the long-term consequences of alcohol abuse differ from the long-term effects of your selected drug?

After you understand how the drug works, try to evaluate the personal side of drug use. If you know someone who uses the drug you have chosen, interview them about it. Ask them to describe its effects. Ask them what they know about how the drug works, where it comes from, how it is made, the consequences of habitual use. Then, share with them what you know about these questions, and incorporate their reactions into your essay.

As a final component of your essay, you may wish to use the infomation you have gathered, both personal and scientific, to argue for a particular public policy regarding the drug. Should it be banned entirely? Or perhaps made available in certain cases, but strongly regulated? Or unregulated and freely available to whomever wishes to use it? These questions will take you into a large, complex, and passionate arena for the discussion of drug use. Focus on your understanding of the biological principles involved, particularly as they relate to the mechanism of drug action and the dangers of habitual use, to develop an opinion about the place for this drug in our society.

## REFERENCES

Anonymous. 1991. Powerful pills—Prozac nominated for "1990 molecule of the year." *Science* 250:1642.

Barinaga, M. 1991. Miami vice metabolite. *Science* 250:758.

Cho, A. K. 1990. Ice: A new dosage form of an old drug. *Science* 249:631–634.

Gawin, F. H. 1991. Cocaine addiction: Psychology and neurophysiology. *Science* 251:1580–1586.

Jacobs, B. L. 1987. How hallucinogenic drugs work. *American Scientist* 75:386–392.

Lewin, R. 1989. Big first scored with nerve diseases. *Science* 245:467–468.

Marx, J. 1990. Marjuana receptor gene cloned. *Science* 249:624–626.

Snyder, S. H. 1989. *Brainstorming: The Science and Politics of Opiate Research.* Cambridge: Harvard University Press.

# *Dandelions Versus the Suburban Turf Warrior* 13

On most summer weekends our nation goes to war. In many ways, it's a larger campaign than Desert Storm. It mobilizes a staggering army of people, an armada of malevolent machines, and a huge arsenal of deadly chemicals. The "enemy" in this pitched battle is any living creature that has the temerity to eat, invade, or in any way disrupt the lawns, gardens, trees, and shrubs that surround millions of homes in this country. Slashing, spraying, and whacking their way toward greener lawns, prettier shrubs, and tidier gardens, weekend warriors across the nation collectively use every physical and chemical means at their disposal to repel the invaders and make their homes safe for those plants and animals lucky enough to merit our cultivation and affection.

Is yard work as war a strained metaphor? Maybe. But the parallel is not entirely unfounded, especially for those scientists and citizens who are increasingly concerned about the larger and longer-term consequences of widespread pesticide use. Uncertainty about public health effects, economic efficiency, and ecological damage associated with pesticides has created a contentious and at times emotional debate about the use of these chemicals, especially for ornamental vegetation like lawns and cultivated shrubs. On one hand, recent studies have suggested that long-term exposure to certain pesticides, even at low concentrations, may produce health and environmental complications not readily apparent from the kinds of tests and assays that are used by governmental agencies to determine if these chemicals are appropriate for public use. But on the other hand, the argument has also been made that the risks associated with pesticide use are in fact grossly exaggerated by environmentalists and scientists with ulterior motives, having more to do with social issues than scientific ones. In fact, the claim has been made that most pesticides are far less dangerous than many compounds that occur naturally in the food found in a normal diet.

Your assignment in this chapter is to explore both sides of this issue by reviewing the available evidence and deciding for yourself exactly how great is the threat posed by the use of pesticides and how legitimate is the practice of using these compounds to maintain ornamental plants.

This exercise will cause you to think not only about the biological principles involved in pesticide application, but also about a strongly held sociological imperative: A well-ordered life must have a well-ordered lawn. A recent *Boston Globe* editorial reflected this imperative when it called on President Bush to allow the lawn around the White House to go back to its natural state. Couldn't the time and money currently used to keep the lawn in tip-top shape be better spent on the homeless, or education, or cancer research, or any of a thousand more productive ways than growing grass? And isn't a meadow filled with hundreds, maybe thousands, of different plants and animals as pretty in its own way as a well-tended lawn? Your answers to these questions, of course, are irrelevant—everybody knows that lawn is gonna get mowed!

If you don't think a lawn is an important social signal—and the White House, in addition to being the home of the first lady, first dog, and first cat, is also the location of the nation's first lawn—suggest to a suburban homeowner that he or she allow the grass to return to a meadow. In many places, to do so would be illegal. Zoning laws prohibit it; neighbors will join together to bring the wayward transgressor back to the fold. If the shaggy lawn persists, a great deal of talk will begin about the chaotic or abandoned appearance of the house, about the deteriorating "quality" of the neighborhood, and finally, about that most troublesome of suburban bugaboos, declining property values—all because someone decided not to cut the grass.

Beyond the broad sociological issue of why we keep lawns is the more specific question of how we do it, or more precisely, how we do it while protecting our health and the safety of our environment. When we use chemicals to control pests that attack our lawns and gardens, we buy into a system of checks and safeguards administered primarily by the federal Food and Drug Administration (FDA) and the Environmental Protection Agency (EPA). That system of risk assessment has recently come under attack, both by those in favor of tighter regulations governing pesticide use and by those who advocate less restrictive guidelines. Most people are aware of the fallibility of pesticide risk assessment. One needs to look no further than the infamous chlorinated hydrocarbon DDT to realize just how difficult is the job of judging the safety of a pesticide and how far-reaching can be the consequences of a wrong decision. The widespread use of DDT to control insect crop pests in the years following World War II was met with unbridled enthusiasm due to the spectacular increase in crop yields that resulted from its application. The effect of DDT was viewed by many agronomists as nothing short of miraculous.

Then the bad news began to come in, first as a trickle, then faster, then as a frightening flood ecological problems, all related to unforeseen aspects of DDT use. It became apparent that DDT accumulates in living tissue, and the rate of accumulation is related to position on the food chain. Organisms at the top of the food chain, including humans and predatory birds such as bald eagles and falcons, experience the highest levels of DDT accumulation—

high enough to cause a variety of physiological problems. For predatory birds, the problem was that DDT interfered with eggshell formation, resulting in eggs with shells that break easily. Several species of birds, including the peregrine falcon, were nearly driven to extinction before DDT use was banned in the United States. It also became clear that DDT is a highly persistent compound in the environment, and year after year of DDT application resulted in soils that were highly contaminated. Winds blowing across contaminated fields pick up the DDT-laden dust, transporting the toxic cargo literally around the world. Even the most isolated environment on earth, the Antarctic, experienced an unintended dousing with DDT that persists to this day. The third serious problem with DDT use is that target pests can become resistant to the chemical, greatly limiting its effectiveness.

There is no real argument among scientists that the history of DDT must be read as an alarming precaution about the sensible application of pesticides. Beyond that consensus, scientists disagree on the safety and utility of current pesticide use. On one hand, some environmental scientists claim that pesticides are increasingly counterproductive and their use should be greatly curtailed. In striking contrast, others argue that the history of DDT and other chlorinated hydrocarbons has created a climate of fear about pesticide use that has stifled research and limited agroeconomic growth. Wild exaggerations on the part of environmentalists, so the claim goes, have led to irrational and unreasonable risk assessment policies by the EPA and FDA. Your assignment is to evaluate both sides of this issue and write an essay that summarizes your opinion of the benefits and dangers of pesticide use.

## GETTING STARTED

You may wish to begin your essay with a brief discussion of how a particular pesticide works (on either plants or insects), and why humans and their pets are not harmed by the chemical. What kinds of long-term environmental consequences could result from the use of this compound? How might nontarget organisms suffer through this compound's application? How has the EPA or FDA tested it in order to ensure that it will not adversely affect public health or environmental safety? After you have reviewed the data, describe your view of the process by which pesticides are sanctioned for public use. Is it excessively restrictive or too lenient? What is your opinion regarding the use of the compound you describe in your essay? Do you agree or disagree with current application practices? What changes would you make in the policies governing its use?

Finally, consider the sociological forces at work that promote or inhibit the ornamental use of pesticides. How easy are these to change? Can we justify the use of pesticides in this way? In your mind, is the ornamental use of pesticides any different from agricultural uses? In answering these questions and developing your opinions, be sure to reference appropriate sources of information in support of your position.

## REFERENCES

Anonymous. 1991. The case against crop chemicals. *Science* 251:517.

Dorfman, A., and J. Leviton. 1991. Can lawns be justified? *Time,* June 3, pp. 63–64.

Marx, J. 1990. Animal carcinogen testing challenged. *Science* 250:743–745.

Ragsdale, N. N., and D. Pimentel. 1991. Letters to the editor: Pesticide use and response. *Science* 252:358.

ReVelle, P., and C. ReVelle. 1988. *The Environment: Issues and Choices for Society.* Boston: Jones and Bartlett Publishers.

Roberts, L. 1991. Dioxin risks revisited. *Science* 251:624–626.

# *Beyond the Limit: Exercise Physiology and Human Athletic Performance* 14

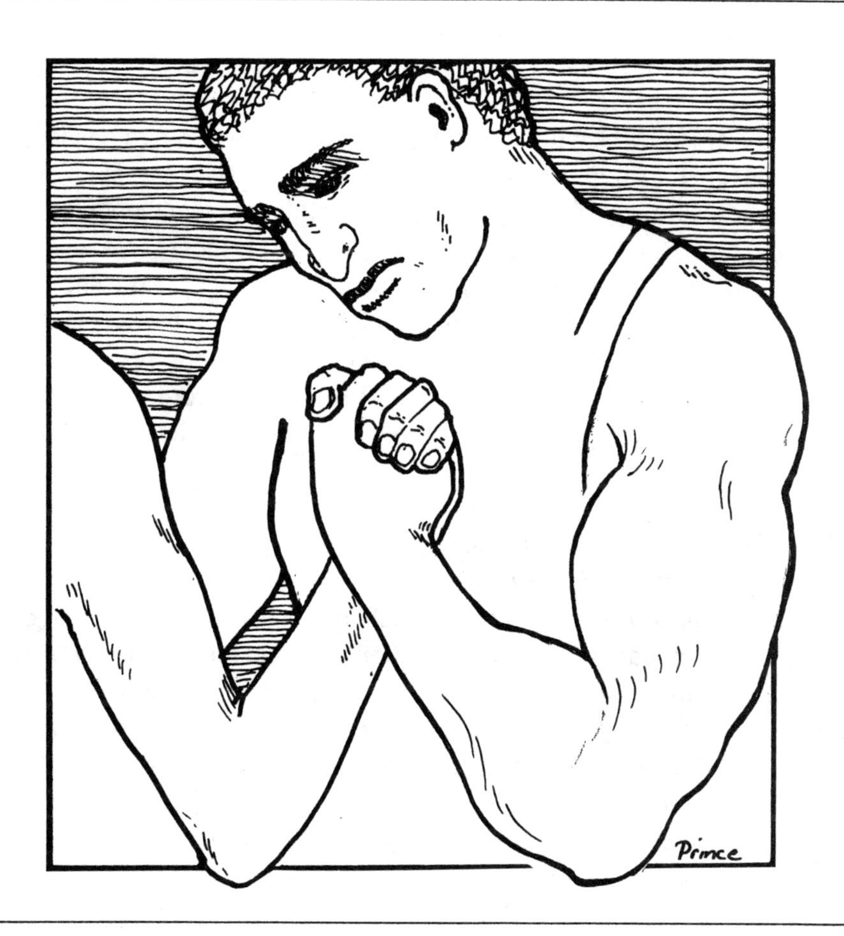


In 1988 Ben Johnson beat Carl Lewis to capture the 100-meter Olympic gold medal. In many people's minds, it was one of the greatest sprints ever. Johnson ran the 100 meters in world record time, shattering the record Lewis had set the previous year. The image of the race remains with many sports enthusiasts for several reasons. First, Johnson was an incredibly dynamic athlete. His body was remarkably muscled. His chest, unusually large for a sprinter, seemed almost to burst with energy. Pacing at the starting line, he looked more like a football player than a world class sprinter. The second intriguing issue surrounding the race was that Johnson and Lewis did not like one another. That was clear from the outset. They had made comments about each other's performances through the newspapers. At the starting line they appeared to deliberately keep their eyes away from one another. In the final 15 meters, when it was clear Johnson had won, he pumped his hand and turned to taunt Lewis.

The final element, which did not surface until a day or two had passed, was that Johnson had apparently been on a regimen of steroids. A routine urinalysis detected traces of anabolic steroids. Johnson, although naturally muscular, had been made even more so by his steady ingestion and injection of highly potent drugs. Naturally a furor erupted. Television showed us various clips of Johnson being hounded by reporters in the Rome airport. Canadians, so proud of their countryman a few days before, now vilified him. The fastest man ever to run a 100-meter race (his record time has since been broken) had only won, so it was reported, because he had resorted to drugs.

In the ensuing weeks, some interesting details surfaced. For one thing, it became apparent that any number of people in the track and field community suspected Johnson had been on steroids for some time. As a result of the steroids, which can affect liver function, the whites of his eyes were jaundiced. The muscle growth he had sustained, according to many observers, was well beyond what could be accomplished by even the most rigorous training. He had become too big too quickly, people said. In addition, he had become somewhat brooding and violent. Some track and field insiders claimed Johnson was running and training with a newly formed aggressiveness.

On a personal level the costs to Johnson, as a result of being discovered as a steroid user, were enormous. He was forced to give back his gold medal, thereby making Lewis the gold medal winner; several of the sanctioning bodies for track and field banned him from competing in future events; the Canadian Olympic committee worded a strong condemnation of his behavior and stated that it would be more vigilant in the future concerning potential drug abuse. Financially, the damage was almost incalculable. Although recently he has returned to racing—and has been offered a million dollars to run a match race against Lewis—the endorsement money he might have earned was suddenly swept away. Instead of being the world's fastest man, Johnson had become a sports pariah.

Johnson's embarrassment was only the first of many recent revelations about athletes using drugs to enhance their performances. The National Football League's Lyle Alzado recently posed for the cover of *Sports Illustrated* with a bandana around his head. The bandana was used to cover his baldness. Next to the picture was a simple statement: "I lied." In the related article Lyle Alzado, an extremely successful NFL defensive lineman for many years, claimed that his use of steroids throughout his career had given him inoperable brain cancer. Moreover, he went on to state that steroids are everywhere in the NFL. According to him, it would be impossible to compete with the huge men who make up the interior lines of most NFL teams if one did not artificially enhance his strength.

At this point most athletic institutions are holding their collective breaths about the steroid question. Certainly major college football players are placed in a bind: If the person lining up against you has been using steroids, then how effectively can you compete without using them? College coaches want stronger, bigger, more aggressive athletes, but at the same time mouth protests against drug use. Yet many former players contend that coaches turn a deaf ear and a blind eye to steroid use among their athletes. Added to this mix is the potential—both for the individual players and coaches and for the institutions themselves—for a stunning financial payoff.

To anyone who follows sports, none of this is news. What might be news, however, is that steroid use is only one way that athletes use to enhance their performances. For many years East German and Soviet endurance athletes have been using a technique called "blood doping." Several months before an important competition is scheduled, blood is removed and the red blood cells are extracted. Just before the event, these cells are reinjected into the circulatory system of the athlete, reportedly increasing the capacity of the blood to transport oxygen and enhancing their performance in athletic events that require high levels of aerobic exercise.

All of this, naturally, raises some important ethical questions. What right do we have as a society to dictate what an athlete can or cannot do to enhance his or her athletic performance? Do we have a responsibility to protect an athlete from harming his or her health? How can we justify our stance against drug use when many of our largest sporting events are underwritten by Budweiser and Miller Beer? If we as a society of sports fanatics want bigger and faster players, and more and more sports of higher and higher quality delivered to our televisions, aren't we inviting substance abuse? Do we merely pretend to care if an athlete ruins his liver by ingesting steroids? If we can tolerate boxing, which is unequivocally harmful to a boxer's health, why do we become so fussy and self-righteous when it comes to drugs?

## GETTING STARTED

This topic may get at the root of what we think of athletics and why we find them so appealing. As Americans, our appetite for athletics seems insatiable. Monday Night Football is an institution in this country. People go out to bars, get together at each other's houses, and sometimes bet enormous sums of money, all to be entertained by the Eagles vs. Packers, or Bears vs. Patriots. Yet despite the fact that American athletics are essentially one long advertisement for various products—beer, razor blades, tires, cars—we still cling to a myth about the purity of the athlete. We apparently want him or her to embody some vague notions of clean living, dedication, and endless training. We are offended when we learn, as is happening with greater and greater frequency, that this or that athlete benefited by using chemical stimulants. Something about it doesn't sit right. It doesn't sit comfortably with the solemn playing of the national anthem at the beginning of all our athletic events. We prefer not to watch the camera pan down the row of waiting football players as they sing, "O say can you see," while imagining them sticking needles in their thighs and buttocks.

In order to approach this essay, you need to give some critical attention to what your attitudes are concerning sports. Where do we draw the line? Aren't athletes who train at higher altitudes "doping" their blood? How do we decide what's legal (or fair) and what's illegal (or unfair)? Ultimately, you may concern yourself with the overiding question: Why do we care? Isn't it up to the individual to supervise his or her own training?

Be careful. Any attack on sports sometimes elicits hasty or unreasoned responses from people running to their defense. Examine carefully what you think about this question. It is sure to be a question that does not disappear from our national attention.

## REFERENCES

Use as a general reference any good textbook on exercise physiology to provide background material for your essay. Then, go back and read the newspaper accounts that were published at the time of Ben Johnson's suspension. The language of the editorials, in particular, might give you an insight in the volatile nature of the issue. Furthermore, *Sports Illustrated* has published any number of articles on steroid use in sports. You may find it helpful to go to the *Science Citation Index* (see Chapter 16) and look up "anabolic steroids" as a topic. You will find a variety of listings.

# *Unit IV The Real Deal: Preparing a Scientific Manuscript*

This unit is about the main event of science writing—the preparation of a scientific research manuscript. Manuscript composition is generally regarded as the single most important challenge to a scientist's communication skills, and the final chapters in this book are dedicated to guiding you through this critical process. The goal of any scientific research report can be simply stated: to communicate in a clear, unambiguous way the outcome and the ramifications of a test of some specified hypothesis. However, for the uninitiated, the apparent path to that goal can be daunting to the point of overwhelming. Part of the reason that scientific writing has this paralyzing effect is that a lot of scientific literature is really hard to read, even for readers with specialized background knowledge and the appropriate technical vocabulary. Readers usually assume that the reason for their difficulty is the sheer intellectual weight of complex scientific concepts and the attending data. In reality, culpability often lies not with the reader, but rather with the writer. The unflattering truth is that many scientific papers are hard to read because they are poorly written.

The four chapters in this unit are structured in a way that will help you to understand the broad intellectual basis of scientific research publication, and the skills, both mechanical and rhetorical, required for assembling a manuscript. The first and *sine qua non* requirement for writing a scientific research paper is that you need a hypothesis—more precisely, a testable hypothesis. Chapter 15 offers some suggestions for helping you to organize your thoughts and develop a hypothesis for your manuscript. If you already have a hypothesis, either provided by your instructor or emerging from an assigned laboratory exercise, then you will not need the suggestions in Chapter 15.

After you have identified a hypothesis, you will test it by designing an appropriate experiment or by making relevant observations. The data you collect will need to be statistically analyzed, and we have provided several simple statistical tests in Chapter 17 to assist you in that analysis. These tests will provide you with a basis for interpreting your results and evaluating your hypothesis, but you will also need to explore the scientific literature for additional information regarding the hypothesis under consideration. To help you do this, we have included a chapter (16) that presents some guidelines for tackling the scientific literature and using it to strengthen the arguments and interpretations presented in your manuscript.

At this point, your manuscript is actually not a manuscript at all—it's simply an assemblage of data and ideas, significant to no one but you. Every scientist knows that unpublished data may give you a warm tingle, but nobody else notices or cares. The final step, then, is to forge from your ideas and data a document that clearly and unambiguously makes a case for your interpretation. Chapter 18 is included to help you do this, both in terms of formating the manuscript, and in terms of enhancing the effectiveness of your written communication.

Creating a scientific manuscript takes time, energy, and commitment. If you are like most scientists, you will have occasions when you struggle with one or maybe all of the stages outlined above, perhaps even to the point of impugning either the character or the ancestry of your humble authors. Don't let these frustrations derail you from the task of developing your ideas to the fullest extent possible. The demands may be high, but so too is the personal satisfaction of producing an effective expression of your scientific analysis.

# *Prospecting for a Prospectus: In Search of a Testable Hypothesis* 15

If you were to ask a dozen scientists to describe a "good scientific hypothesis," you would probably get a dozen different answers. Every scientist has his or her own favorite example of a hypothesis that has proven to be illuminating and useful far beyond the original vision. But despite the differences in specifics, each of the scientific hypotheses would share one simple and defining characteristic: They would all be unambiguously testable. In other words, these hypotheses can be falsified (i.e., shown to be wrong) by the appropriate experiments or quantitative observations. Sometimes that feature of a hypothesis is not so obvious. Sometimes an idea is so clever, interesting, or enlightening that the absence of "falsifiability" is lost in the background. Such ideas may be fun and entertaining and even dynamically important in the affairs of people, but if they can't be fully tested, then they fall outside the realm of science.

Consider, for example, a hypothesis about one of the oldest and most spectacular mysteries in the history of the organic world: the sudden and inexplicable disappearance of the dinosaurs approximately 65 million years ago. Countless explanations for this dramatic event have been published, ranging from the bizarre to the plausible. Among the more whimsical is an argument that dinosaurs became extinct because their gonads were unable to produce viable sperm due to the debilitating effect of changing climatic conditions. According to this account, dinosaur populations crashed 65 million years ago because eggs went unfertilized when male dinosaurs were rendered sterile by rising environmental temperatures. In a nutshell, *Tyrannosaurus rex* was shooting blanks. The idea is based on more than just conjecture. In fact it is based on observations of living species of vertebrates (including humans) in which testicular dysfunction can result from elevated temperatures during spermatogenesis. Although this explanation may have a certain appeal to resourceful thinking, it lacks that most basic of features: It cannot truly be tested. We will never understand the reproductive physiology and anatomy of male dinosaurs, given the dearth of living specimens and the absence of fossilized dinosaur testicles (soft tissues such as the heart, liver, digestive tract, etc., don't fossilize). We can never conduct the experiments or make the measurements necessary to test the predictions of this hypothesis. Whatever the merits of this idea may be, it will be forever relegated to an area outside of science.

## GETTING STARTED

With this important caution about scientific hypotheses in mind, the purpose of this chapter is to introduce you to relatively simple projects that can serve as the basis for preparing a scientific research paper. We hope this exercise will enable you to improve your writing skills as well as provide you with an introduction to scientific research. To complete the assignment, you will need to formulate a hypothesis that can be tested by original quantitative observations. Most of the following topics deal with some aspect of ecology, due primarily to the relative simplicity involved in collecting the appropriate data. Your instructor may have additional suggestions for topics in other fields.

1. The relationship between prey density and feeding rate in aquarium fish, including an analysis of inter- and intra-specific differences in fish behavior.
2. Distribution of mosses and lichens on tree trunks (vertical distribution, species of trees, geographical orientation).
3. Flower morphology—what are the characteristics of early spring flowering plants? How are they distributed through the habitat?
4. Functional significance of tree shape and its relationship to tensile strength of wood, tree height, habitat, etc.
5. Functional significance of leaf shape and its relationship to sunlight intensity, tree size, habitat, etc.
6. Territoriality of domestic cats, pigeons, or some other domestic animal to which you have access and can observe.
7. Effect of current on attached algae, or the effect of current on habitat use by aquatic insect larvae in a local stream.
8. Behavior, body size, beak morphology, foraging efficiency, etc., in birds and how they relate to the type of food used (seed size, suet, fruit type, etc.).
9. The importance of feeder height and food type on bird species composition at a bird feeder.
10. Community composition in an ephemeral habitat such as a rain puddle, hollow of a tree, tidal pool, summer pond, etc.
11. If you can get to the coast, differences in the rocky intertidal community with elevation (duration of exposure). Distribution of barnacles in relation to the micro-topographic features (crevices, rocks, etc.).
12. Effects of seed size, shape, ornamentation, fruit type, etc. on patterns of seed dispersal.
13. Differences in species composition, diversity, and growth among groups of related habitats: lawns, cemeteries, pastures, roadsides.
14. Foraging behavior in ant colonies. How far do ants forage from the nest? What types of food do they collect?
15. Change in vegetation, e.g., plant species, as distance from water increases up a slope.

Be aware that these topics are intended to get you thinking; it is up to you to develop one of them into a research project centered around a testable hypothesis. After you have discussed your research project with your instructor, prepare a one-page description (your prospectus) of what you propose to study, how you will test your hypothesis, what you hope to find out, and why you think this information will be of interest to a broader audience. You will need to choose your words carefully to cover these topics in a one-page essay.

After your prospectus has been approved, work with your instructor to design and carry out the experiments or observations necessary to test your hypothesis. The data you collect can be analyzed using the simple statistics presented in Chapter 17; if more complex analyses are required, consult with your instructor. Upon completion of your data analysis, you will want to explore the scientific literature for additional information about your project. Use Chapter 16 to help identify and find the latest published sources.

When you are ready to start writing your manuscript, refer to Chapter 18 for suggestions regarding organization, format, and style. Chapter 18 also suggests ways of getting started, organizing your time, and editing your work that may prove helpful in composing your manuscript.

# *Using and Abusing the Scientific Literature* 16

One of the early ideas for the artwork in this chapter was to have a bewildered, barely visible scholar seated at a table, surrounded by enormous, tottering stacks of books, with still more books looming overhead, poised precipitously on row after row after row of shelving. The entire image was supposed to create the impression that a live, perhaps bloody, burial was at hand, the erstwhile scholar done in by the sheer volume of an overwhelming scientific literature. Fortunately, the sensibilities of the artist prevailed over the strained melodramatic vision of the authors, and the "human sacrifice" concept was dumped in favor of an image that reduces the implied threat of the literature while enhancing the composure (and presumably the longevity) of the scholar.

Even though the idea was scrapped, it is worth recounting here because it conveys an important point about the scientific literature: The burgeoning volume of current scientific publications can be overwhelming almost to the point of paralysis for the uninitiated. Every year in biology alone, hundreds of thousands of new papers are published in journals from countries around the world. New, increasingly specialized journals seem to spring up almost overnight, their presence announced in every batch of scientific junk mail shuffled across a professor's desk. "Staying current" (an expression used to describe familiarity with the most recent scientific advances) is becoming possible only by focusing on a small, narrowly defined field. Many research laboratories now assign full-time staff personnel the job of scanning the literature for relevant papers. If keeping track of research publications is a monumental job for professional biologists, what hope can there be for an undergraduate struggling to find some information for a *Biology Write Now!* writing assignment?

The answer to that question is "plenty," due primarily to a second critical feature of contemporary scientific publications: the electronic transfer of information and the use of computers. These make it possible to explore and use the literature in ways unimagined just a few years ago. It is now possible to track subjects, ideas, techniques, and authors through the literature quickly and easily using a variety of indexing and abstracting sources. These sources differ in the services they offer. The following brief descriptions of three widely available sources of this type will help you understand how they differ and how you can use them in developing the background material for your writing assignment.

## SCIENCE CITATION INDEX

This indexing source has emerged as an extremely powerful tool for tracking down information in the scientific literature. Using *SCI* is a relatively simple affair. References are accumulated throughout the year and published in three interrelated portions: a source index, a subject index, and a citation index. The source index lists alphabetically by author every scientific paper published in a calendar year. The subject index organizes those publications by subject matter, using key words taken from the titles. The citation index lists every paper alphabetically by author that was *cited* (not published) in any article appearing in the source index. Visually, all this can be somewhat imposing because a single year of *SCI* may take up several feet of shelf space in the library. But, this index is actually quite easy to use and makes it possible to track the progress of an article or an author through time, right up to publications within the past few weeks. Suppose, for example, you have read a paper that was published by Smith in 1981 and want to know what happened to the ideas presented therein over the ensuing years. By using *SCI* it is possible to find out who cited that paper since its publication, where the citations occurred, and what is the current status of the ideas. In other words, it doesn't take much, an author's name from a paper several years ago, even a vague knowledge of a subject, and you can crowbar into the literature to discover a mountain of relevant information about the subject of interest.

### BIOLOGICAL ABSTRACTS

The *SCI* will help you discover relevant references for a given subject, but it will not provide any information about the articles beyond the titles, dates of publication, and scientific journals in which they were published. *Biological Abstracts* provides the abstract (i.e., a thumbnail sketch of a paper's content) for each of nearly 200,000 biological and biomedical research papers published in more than 8,000 journals every year. This service is especially valuable if you need to know more about a paper that appeared in a journal or periodical that your library doesn't have. The abstracts are indexed in several different ways: by author, by taxonomic categories, by subject, etc. In contrast to *SCI,* these indexes are not cross-referenced from year to year, nor do they include information about citations appearing in the papers.

### ZOOLOGICAL RECORD

*Zoological Record* is an annually compiled, taxonomically based compendium of the zoological literature published worldwide. It is organized taxonomically in 27 different sections, 25 of which deal with different animal groups. The two remaining sections deal with new generic and species names appearing in the other sections, and with general zoological literature. Within each section, there are five indexes: the author index, the subject index, the geographical index, the paleontological index, and the systematic index. The indexes give a brief indication of the important aspect of every paper listed. In the geographical index, articles are assembled by geographical area, allowing one to find all the papers published in a given year for a particular group of animals from a particular place. Likewise, the paleontological index sorts articles by geological eras. As with the *Biological Abstracts,* the *Zoological Record* is not cross-referenced from year to year.

## OTHER SOURCES OF INFORMATION

The focus of this chapter thus far has been on what is called the primary scientific literature, that is, original research papers appearing in scientific journals. For many of the writing exercises in the first three units, more general references may be of greater use to you than research papers. A good place to start looking for these is the "Additional references" list found at the end of the chapters in any good introductory biology textbook. After you have checked into these, you may wish to use sources such as those indicated above to delve further into the literature.

Then again, you may not. As was indicated in the "To the Student" section at the beginning of this book, don't be afraid to explore alternatives to the standard sources of scientific information. Television programs, radio shows, articles in *Sports Illustrated,* accounts in the newspaper, even relevant personal communications from someone with direct knowledge or personal involvement in the issue being considered, any of these may be of use to you, not only for factual information, but also to get you thinking about the issues. If you do rely on alternative sources of information, be sure to report it in a coherent way in the literature cited section of your paper.

## ABUSING THE LITERATURE

And finally, a word about plagiarism: DON'T. Don't do it. Don't let someone else do it with your work. A common mistake made by students just learning how to write scientific papers is to assume that immunization against charges of plagiarism is conferred simply by inserting the true author's name at the end of long passages, even whole paragraphs, of word-for-word transcription of the original text. If they aren't your words, you have no right to them. Lifting whole blocks of text without accompanying quotation marks is plagiarism, regardless of whether the author's name is indicated after the passage or not.

# *Number Crunching with Simple Statistics* 17

If you're lucky, you'll be able to skip most of this chapter. If the results of your experiment or the patterns of your observations are so obvious that the outcome is clear to everybody, then statistical analysis becomes irrelevant. In fact, most scientists view statistics as important only when the results aren't obvious, when it isn't so clear exactly what the outcome is, or if it's consistent with the hypothesis being evaluated. For example, if you are testing the effectiveness of a drug to cure the common cold and find that every patient responds immediately when they use it, you don't need a statistician to tell you that you're on to something big. If only half the patients respond, you may still feel pretty confident that the drug has an effect, but now you wonder about what happened to the nonresponsive patients. And if only one or two out of every ten patients respond, you might begin to think that the drug is generally useless, although it appears to have a puzzling effect on a small number of people. Maybe the few responsive patients would have gotten better anyway, just by chance, and the drug has no real effect, but you don't feel comfortable with that conclusion because it is based solely on intuition and subjective conjecture.

Statisitics offers an objective, unbiased way to evaluate results such as these that fall into the gray area where "yes" and "no" answers give way to "maybe." Perhaps unfortunately, variability in the results of many scientific experiments pushes them into that gray area, and statistics have become an indispensible tool for teasing apart the variation in order to determine if the results conform to the predictions of the hypothesis under consideration. This chapter is intended to introduce you to some basic statistical techniques that are commonly used to analyze the variablity found in experimental results.

## Sampling Techniques

Many biological experiments involve making measurements on only a small subset of the total population under study. In order for statistical analyses to be valid, sampling must be carried out in a random manner. A random sampling procedure simply means that every individual in the population has an equal chance of being chosen. The selection of individuals for inclusion in a study or for assignment to an experimental group should be done randomly. Your instructor can show you ways to do this.

## Summarizing the Data

Many types of data are best summarized by calculating the mean and the variance of the sample. The mean (represented as $\overline{x}$) is the arithmetic average of the individual sample values (x). The variance ($s^2$) is an estimate of the variability among values within the sample and is based on the deviations of individual values from the sample mean. The larger the variance, the greater the dispersion of individual values around the sample mean. Another measure of variability is the standard deviation (SD), which is simply the square root of the variance. The variance can be calculated as the sum of the squared deviations of individual values from the mean, divided by one less than the sample size (n). In equation form:

$$s^2 = \frac{\Sigma(X - \bar{X})^2}{n-1}$$

The mean of a sample is actually only an estimate of the true population mean, unless, of course, every individual in the population has been included in the sample. In order to estimate the accuracy of the sample mean, both the amount of variation in the sample and the size of the sample must be taken into account. These two factors are incorporated into a value called the standard error (SE) of the mean, and this value is calculated as follows:

$$SE = \sqrt{\frac{s^2}{n}}$$

Obviously, if the sample size is very large, then the standard error becomes very small, indicating a high level of confidence that the sample mean is very close to the true population mean. Conversely, if the data are highly variable and the sample size is small, then the standard error is large and our confidence in the sample mean as an accurate reflection of the true mean is relatively low.

## COMPARING MEANS

The analysis of experimental results often centers on a comparison of mean values from two or more different samples. Suppose you wished to know, for example, if the frogs found at a pond near a nuclear waste dump were smaller (or larger) than those located elsewhere. You would measure the body size of animals in a sample taken from the vicinity of the dump and compare that value with the mean body size in a sample taken from an area some distance away. A comparison of this type must account not only for the difference between the means, but also for the variation among individuals within each sample. For example, a half inch difference in the average size of the frogs in the two samples would mean very little if both populations included many individuals that varied in length by several inches. On the other hand, that same half inch difference would be highly significant if all the frogs collected in the two samples were within a quarter inch of the sample mean. In both cases, the central question is the same: How likely is it that the half inch difference between samples arose purely by chance? Statistical analysis can provide an answer to that question.

One method for comparing means that takes into account these considerations is called the "*t*-test." In this test, a "*t* value" is calculated that allows one to evaluate the difference between the means, relative to the variances of the samples. The following formula indicates how a *t* value is calculated:

$$t = \frac{\bar{X}_1 - \bar{X}_2}{\sqrt{SE_1 + SE_2}}$$

The *t* value calculated in this way is used to test if the difference between means is "significant." Statistical significance is a way of saying that the likelihood (or probability) of the observed difference in means occurring entirely due to chance alone is so small that we can in fact eliminate chance as an explanation for our results and instead attribute the difference to some other factor. This is really the heart of every statistical analysis and bears restating: Statistics provides ways of determining the probability of some event or series of events occurring due to chance alone. For example, we know intuitively that two-children families with one boy and one girl are quite common and that ten-children families with all boys are extremely rare. Statistics allows us to replace intuition with a precise statement of the actual probability of these events occurring due to chance alone.

In fact, "chance alone" forms the basis of an important concept in hypothesis testing, the null hypothesis ($H_0$). The null hypothesis assumes that there is no real difference between samples other than that which could be expected due to random variation in sampling. In the frog example given above, the null hypothesis can be stated in the following way:

$H_0$: Frogs collected from the pond located near the nuclear waste dump do not differ in size from those collected at a pond several miles from the dump.

The null hypothesis is then tested by comparing the calculated *t* value with the values given in the table at the end of this chapter. The table values present the probability of obtaining a particular *t* value even when the null hypothesis is true. The magnitude of these tabled *t* values is affected by the total number of individuals included in the experiment, expressed as the degrees of freedom (*df*). The *df* is calculated as $n_1 + n_2 - 2$. So putting all this together, if you collected ten frogs from one pond and twelve frogs from another pond, you would expect that 95 percent of the time the calculated *t* value would be less than or equal to 2.09 (reading from the table with $df = 20$ and Probability of a larger value = 0.05) if the null hypothesis were true. If the calculated *t* value is greater than the tabled *t* value, then scientists generally interpret this to mean that the observed differences did not occur due to chance but instead due to some other factor. The null hypothesis is then rejected and an alternative hypothesis may be suggested.

## COMPARING FREQUENCIES AND DISTRIBUTIONS

The *t*-test described above deals with evaluating *continuous* variables (i.e., characteristics that vary continuously through a population, such as height or weight in humans). Biological variables may also be organized into *discrete* classes and evaluated on the basis of the relative frequencies of occurrence. Observational data, such as estimates of habitat preference (e.g., mice distributions in open fields versus in mature forests) generally fall into this category. A common method for evaluating data with discrete distributions is called the chi-square ($X^2$) test. As with the *t*-test, the chi-square is used to determine the likelihood that a particular pattern of distribution occurred due to chance alone. It too is based on a null hypothesis, but in the case of the chi-square test, the null hypothesis is used to generate "expected" values that can be used in conjunction with the actual data to calculate the $X^2$ statistic. The following formula shows this calculation:

$$X^2 = \Sigma \frac{(O - E)^2}{E}$$

where O is the observed or experimentally measured value and E is the expected value based on the null hypothesis. As with the *t* value, the calculated $X^2$ value is then compared to a table (at the end of this chapter) of values that would be expected to occur if the null hypothesis were true. The degrees of freedom in the $X^2$ test are determined by the number of categories, minus one. The use of this test is revealed in the following example. Suppose you wanted to evaluate the effect of bird feeder height on the number of birds using the feeder. Your observations consist of watching birds visiting two feeders: one on the ground, the other 10 feet above the ground. The null hypothesis is that height has no effect, and so the number of birds visiting the feeders will be the same. You make your observations and find that a total of 90 birds visited your two feeders—29 of them to the ground feeder, 61 to the other. How likely is it that this pattern of distribution could have occurred due to chance? The following calculations provide an answer:

| | Ground | Aboveground | |
|---|---|---|---|
| Observed | 29 | 61 | |
| Expected | 45 | 45 | |
| O – E | –16 | 16 | |
| $(O-E)^2/E$ | 5.7 | 5.7 | $X^2 = 11.4$, $df = 1$ |

The calculated $X^2$ exceeds the table value shown below with $df = 1$ (there are only two categories of feeders) and a significance level of 0.005. Therefore, you reject the null hypothesis that there is no difference (other than that due to chance) in the number of birds visiting the two feeders.

| Distribution of $t$ if the null hypothesis is true | | | | | |
|---|---|---|---|---|---|
| Degrees of freedom | Probability of a larger value, sign ignored | | | | |
| | 0.050 | 0.025 | 0.010 | 0.005 | 0.001 |
| 8 | 2.31 | 2.75 | 3.36 | 3.83 | 5.04 |
| 9 | 2.26 | 2.69 | 3.25 | 3.69 | 4.78 |
| 10 | 2.23 | 2.63 | 3.17 | 3.58 | 4.59 |
| 11 | 2.20 | 2.59 | 3.11 | 3.50 | 4.44 |
| 12 | 2.18 | 2.56 | 3.06 | 3.43 | 4.32 |
| 13 | 2.16 | 2.53 | 3.01 | 3.37 | 4.22 |
| 14 | 2.15 | 2.51 | 2.98 | 3.33 | 4.14 |
| 15 | 2.13 | 2.49 | 2.95 | 3.29 | 4.07 |
| 16 | 2.12 | 2.47 | 2.92 | 3.25 | 4.02 |
| 17 | 2.11 | 2.46 | 2.90 | 3.22 | 3.97 |
| 18 | 2.10 | 2.45 | 2.88 | 3.20 | 3.92 |
| 19 | 2.09 | 2.43 | 2.86 | 3.17 | 3.88 |
| 20 | 2.09 | 2.42 | 2.85 | 3.15 | 3.85 |

| Distribution of $X^2$ if the null hypothesis is true | | | | | |
|---|---|---|---|---|---|
| Degrees of freedom | Probability of a larger value | | | | |
| | 0.100 | 0.050 | 0.025 | 0.010 | 0.005 |
| 1 | 2.71 | 3.84 | 5.02 | 6.63 | 7.88 |
| 2 | 4.61 | 5.99 | 7.38 | 9.21 | 10.60 |
| 3 | 6.25 | 7.81 | 9.35 | 11.34 | 12.84 |
| 4 | 7.78 | 9.49 | 11.14 | 13.28 | 14.86 |
| 5 | 9.24 | 11.07 | 12.83 | 15.09 | 16.75 |
| 6 | 10.64 | 12.59 | 14.45 | 16.81 | 18.55 |
| 7 | 12.02 | 14.07 | 16.01 | 18.48 | 20.28 |

# Effective Scientific Writing: Style and Format

# 18

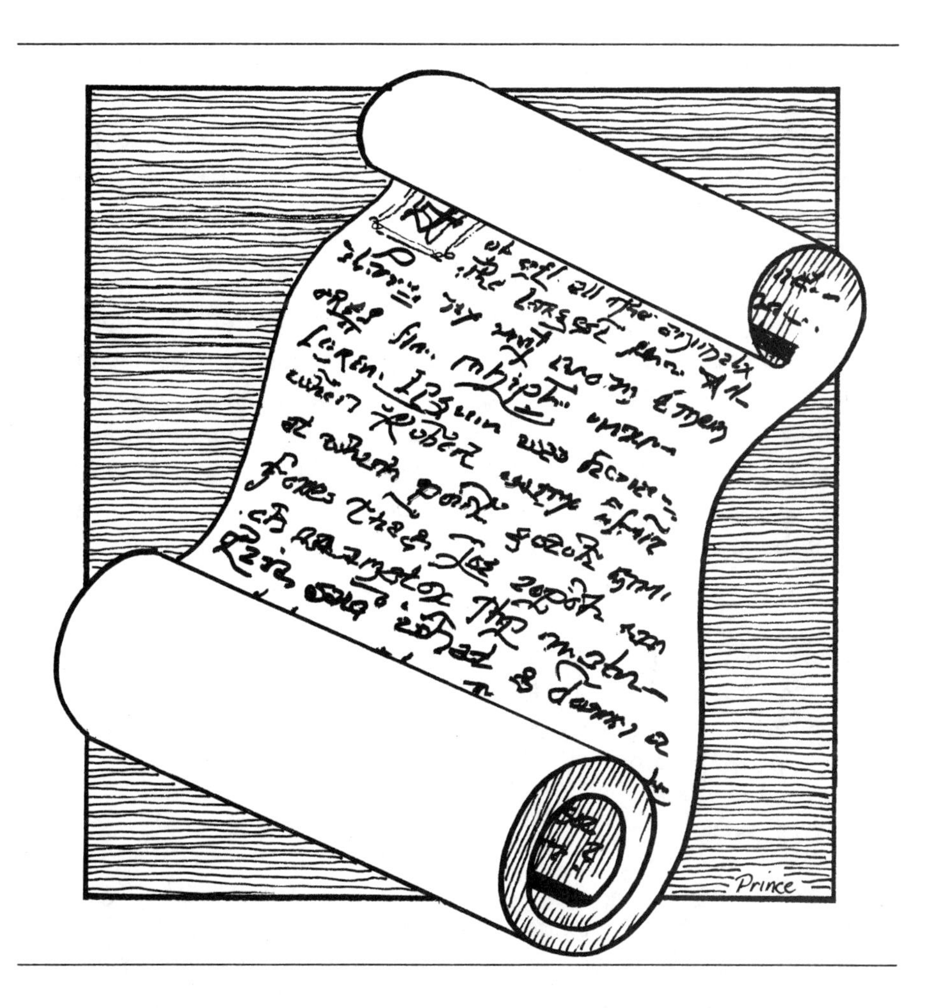


Scientists have a difficult job when it comes to reporting their findings. Frequently the material is dense. It is often so intricately developed, with one element resting on the next, that sentences can quickly groan under the weight. Generally the advice of editors to writers who have scribbled cumbersome sentences is to break the sentence down. Chop it up. Distinguish the parts. Write plain, declarative sentences.

Scientists, due to the demands of their material, are not always able to follow this advice, however. Often the material—statistical analyses, for example, or descriptions of complex cellular biochemistry—requires complex language. In addition, because scientists often use terms and phrases unfamiliar to people outside the scientific community, their language can quickly border on the arcane. As a result, scientific reports occasionally sound almost like a different language. Read an extremely technical paragraph on recombinant DNA, for instance, and, if you are not well versed in the subject, you will soon find yourself tearing out your hair.

Despite the complex material, however, scientists have the same requirement as everybody else to understand the power and structure of language if they are to write effectively. Regardless of how technical a subject may be, clear writing can help you in describing your results and supporting your interpretations. Any scientist who ignores the forms and shapes of language, does so at his or her peril. Complex does not mean muddy. Difficult does not mean poorly worded. Important technical details will become be lost, or even worse, annoyingly incomprehensible, if they are described in weak, flabby language.

In order to begin thinking about your writing assignments in this book, you should consider some of the following topics. Not all of them will apply in every exercise, but most will come in handy at one time or another in your writing career.

## The Audience

One of the first questions a professional writer will ask when given an assignment is: Who is going to read it? Writing for *Scientific American*, a general interest publication widely distributed among scientists and the public, is very different from writing about the same topic in the *Journal of Experimental Biology*. The essays in this book—and probably in the initial stages of your academic career—will be pointed toward an audience consisting of your instructors and classmates. What then, ask yourself, can you assume about your audience? What type of language is appropriate for academic discourse? What's the best way to approach a writing assignment given in a biology course?

Intuitively, you already know the answers to some of these questions. You're not going to use slang, for example, or coarse language. Such language trivializes the whole writing exercise and would be perceived as an insult by your instructor. At the same time, be careful not to become too formal. Formal language can be stilted and distracting. Unless you usually speak in a fairly formal style, it is unlikely that you should write the following sentence: "The propinquity of the river was a propitious factor in the establishment of a large bald eagle population." Your language will sound artificial and unconvincing if you try to adopt a transparently pretentious voice. Aim for a firm, comfortable voice. Aim for a style that does not sound condescending or club readers over the head with your superior knowledge. Aim for a style that invites readers into an exchange of ideas.

You are fortunate because you are writing to an audience that already grasps most of the material you will present. In most cases you do not have to explain all the terms at great length, since your instructors will already understand them. In some ways, this makes your writing assignments easier since you don't have to encumber your text with burdensome details. In other ways, this makes your assignment a much more subtle one, since you must work carefully to reach the right mix of background material and assumed knowledge.

## Spelling, Grammar, and Other Headaches

In this day of computerization, there is no excuse for misspelled words. SPELL CHECK EVERYTHING! If the spell checker does not have a suggestion on a misspelled word, then go to a good dictionary. Do not shrug your shoulders and swallow deeply, hoping the word is about right. Good spelling is not a virtue; poor spelling is not a crime. Laziness is one of the seven deadly sins, however, and poor spelling is just plain laziness. The fact that you are in college should mean spelling is no longer an issue. It simply will be correct. Your instructor will expect no less.

Grammar, naturally, is a much more complex issue. It is more complex, to begin with, because research indicates that most of our grammatical patterns are established by the age of five. You can change. You can improve your communication skills. Over long periods you can even adapt to unfamiliar patterns. But generally, you write in a rough approximation of how you spoke at the age of five. The words become more sophisticated, but the language structure remains consistent. Obviously, then, this is not the place to try and overhaul your grammar skills. If, however, you can begin to think of language as a framework, an elastic understructure for your ideas and perceptions, then you will be on the road to appreciating the strengths and pitfalls of language use. To illustrate a small example concerning language structure, consider the two sentences below.

> Little attention is being paid to northern boreal forest communities by avian ecologists.
>
> Avian ecologists are paying little attention to northern boreal forest communities.

The first sentence is called passive writing. The second is called active writing. In scientific writing, passive voice is typically used to communicate an objective statement of the results or the experimental conditions. The active voice is more direct and dynamic, and is often used for the narrative portions of a manuscript or essay. However, be aware that dogmatic and unrelenting use of either active or passive voice in your writing can become tiresome and ponderous. Mix it up, in a way that will keep your reader moving with you from one idea to the next.

One of the most serious distractions in scientific writing is the separation of the subject of a sentence from its verb. The longer we must wait to find the verb in a sentence, or the greater the distance between the subject and the verb, the more uneasy we become to know who did what. This observation notwithstanding, some scientific writers seem intent on stuffing an inordinate number of adjectives, adverbs, qualifying phrases, and intervening asides between the subject and the verb. The impact of the sentence and the importance of the conveyed information is often completely diffused as the reader waits and waits for the verb to come wandering by. Verbs should follow subjects as soon as possible, meeting the expectations of the reader and eliminating unnecessary distractions.

## Using Anecdote and Drawing Analogy

One of the secrets professional writers learn early on is the value of anecdotes and analogies to liven up their essays. Anecdote allows us to show rather than tell. Telling, most writers understand, quickly becomes a drone. Showing, on the other hand, is one of the primary reasons we read. Leo Tolstoy, the great Russian novelist, claimed all literature is simply an expression of our desire to peep through the keyhole of various rooms.

In writing a scientific manuscript, the use of anecdotes is much more limited, primarily because interpretations and explanations of the data must be based on hard evidence, rather than anecdotal information. That said, however, be aware of the power of an image in conveying information. Again, consider two examples below.

> The energy expenditure during a 5-mile jog is about 500 kilocalories for an average person. This "cost" is met by burning a combination of fats and carbohydrates during the run to fuel muscle activity. The relative contributions of these two substrates varies from person to person, depending on their diet and physical condition.

> The energy expenditure during a 5-mile jog is about 500 kilocalories for an average person. Though a 5-mile run is fairly strenuous exercise, this "cost" is actually less than the energy contained in a typical hamburger sold at a fast-food restaurant. If fat were the only substrate burned to support muscle activity while running, the average person would burn less than one half of an ounce of fat for every mile run. Most people burn a combination of fat and carbohydrates during exercise, depending on their diet and physical condition.

It is difficult to grasp what 500 kilocalories means to us. We've never seen 500 kilocalories, so how can we visualize it in any meaningful way? The second example shows us how. By relating it to the energy in a hamburger or a couple of ounces of fat, we immediately have a pretty good notion not only about the energetics of exercise, but also about the high energy content of fat. We see the cost rather than hear it.

Obviously not all science writing lends itself to imagery and anecdote, but a good image can save a teetering essay. And a corollary of that observation is that a single clumsy or inappropriate analogy can, perhaps unfairly, completely deflate a paper. You should look for opportunities to use images, anecdotes, and analogies to clarify your writing and make it more engaging.

But also be aware of the importance of clear thinking and logical construction in drawing your images.

## FINAL WORDS ON STYLE

One of the surprises you'll find as you write more is that you cannot escape your voice. You write as you talk. If you need verification of this phenomenon, think about letters you receive from friends. One of the reasons they can make you smile is that they carry with them your friend's voice. Depending on how skilled your friend might be at writing, his or her voice will come through even more clearly.

Voice is an appropriate word since it relates writing to singing. Children, if you notice, are often nearly tone deaf. They can't carry a tune; they can't hit a soothing note. Gradually, however, they learn to sing. Many become accomplished. They learn to control their voices. They get their voices to express, musically, whatever it is they ask of it. Science writing will challenge you with a new style of music. To meet the challenge and sing with it, you must first develop and then have confidence in your own voice.

Don't be afraid to find your own personality in your writing, and don't think of style and grace in scientific writing as something different from those features in other types of prose. Good writing is good writing, whether it be found in the pages of a scientific journal or a pulp fiction novel. Above all, avoid buying into the common prejudice that science writing, by definition, must be dry and ponderous.

## FORMAT

The following format is suggested when you prepare your scientific manuscript. When a scientist actually begins to prepare a manuscript for publication, the format used depends on the scientific journal. Each journal has a particular style for presenting the paper. This style is outlined in the "Instructions to Authors" found in either the front or back of each volume of the journal. Check with your instructor to see if you are required to follow the format of a particular journal. If not, use the following. The literature cited format given here should also be used when citing sources in the essays from the first three units.

**Title:** A title should be brief, specific, and descriptive. Where appropriate, it should include the nature of the study, the organisms studied, the technical approach. Avoid very broad, general titles that convey little information. The title page should include the title and also identify the author.

**Abstract:** The abstract summarizes the contents and conclusions of the paper, points out new information in the paper, and indicates the relevance of the work. State briefly and succinctly what the paper reports. Do not make statements such as "Feeding behavior of fish is discussed." Write instead "In the summer, largemouth bass feed most actively between 9:30 and 11:00 A.M."

**Introduction:** The introduction should open with a short statement about the subject, should explain the purpose of the investigation (i.e., the hypothesis "at risk" in the study), and should briefly orient the study to its field. This section usually does not exceed two pages; be brief where possible, but be sure the reader knows what to expect in the rest of the paper.

**Methods and Materials:** Indicate in this section the techniques and equipment used in your experiments or observations in sufficient detail so that another worker can repeat the procedures exactly. You should also include here information on where and when you conducted your study. Unless the equipment is new or unusual, it is not necessary to outline the theory of its operation in detail. Be sure to indicate the number and type of organisms used in your study. It may be convenient to subdivide this section according to study area, experimental techniques, statistical analysis, etc.

**Results:** Outline the results of your experiments, including tables, graphs, and figures where appropriate. Along with the visual presentation of the results, present a descriptive statement of the data. Arrange the data in a unified and coherent sequence so that the manuscript develops clearly and logically. As with Methods and Materials, it is frequently useful to subdivide this section using subheadings. Every table should include an explanatory heading, and every figure should be accompanied by a descriptive caption below the figure. If you conducted a statistical analysis of your results, that should be presented in this section.

**Discussion:** In this section you should interpret your results. Did your observations conform to the predictions of the hypothesis? How did they differ from your expectations? Can you account for unanticipated patterns in your data? Be sure to support your interpretations with references from the literature.

**Literature Cited:** This section should include all the sources that you cited in the text of your manuscript. The proper way to cite a reference in the text of your paper is as follows: "Water balance in desert rodents is strongly influenced by seed selection (Smith, 1990)." or "Smith (1990) has argued that seed selection influences the water balance of desert rodents." Every reference that appears in this section must be mentioned somewhere in the text. The following are examples of how to cite references:

Postlethwait, J. H., and J. L. Hopson. 1992. *The Nature of Life,* 2d. ed. San Francisco: McGraw-Hill, Inc.
Breed, M. D. 1988. Genetics and labour in bees. *Nature* 333:299–301.

The first of these references is the proper form for a book, the second for a scientific journal.

**Final Editorial Notes:** Your manuscript should be double-spaced with a minimum of 1-inch margins on the sides, top, and bottom. The pages should be numbered consecutively, starting with the title page. All units of measure must be metric, but nonmetric units may be presented parenthetically. Underline or italicize all scientific names for species. Names of higher taxa (families, orders, etc.) are not italicized. Explain in full all but the most common acronyms. Don't use long words when a short word will do (e.g., "use" not "utilize"). The word "data" is plural and should be treated as such in the text. Therefore, "The data are consistent. . . . " is the proper construction.

JUDI LINDSLEY